RNA Exosome

ADVANCES IN EXPERIMENTAL MEDICINE AND BIOLOGY

Recent Volumes in this Series

Volume 694
PROTEIN METABOLISM AND HOMEOSTASIS IN AGING
Edited by Nektarios Tavernarakis

Volume 695
THE CELL BIOLOGY OF STEM CELLS
Edited by Eran Meshorer and Kathrin Plath

Volume 696
SOFTWARE TOOLS AND ALGORITHMS FOR BIOLOGICAL SYSTEMS
Edited by Hamid R. Arabnia and Quoc-Nam Tran

Volume 697
HOT TOPICS IN INFECTION AND IMMUNITY IN CHILDREN VII
Edited by Andrew Pollard, Adam Finn and Nigel Curtis

Volume 698
BIO-FARMS FOR NUTRACEUTICALS: FUNCTIONAL FOOD AND SAFETY CONTROL BY BIOSENSORS
Edited by Marisa Teresa Giardi, Guiseppina Rea and Bruno Berra

Volume 699
MCR 2009: PROCEEDINGS OF THE 4TH INTERNATIONAL CONFERENCE ON MULTI-COMPONENT REACTIONS AND RELATED CHEMISTRY, EKATERINBURG, RUSSIA
Edited by Maxim A. Mironov

Volume 700
REGULATION OF MICRORNAS
Edited by Helge Großhans

Volume 701
OXYGEN TRANSPORT TO TISSUE XXXII
Edited by Duane F. Bruley and J.C. LaManna

Volume 702
RNA EXOSOME
Edited by Torben Heick Jensen

A Continuation Order Plan is available for this series. A continuation order will bring delivery of each new volume immediately upon publication. Volumes are billed only upon actual shipment. For further information please contact the publisher.

RNA Exosome

Edited by

Torben Heick Jensen, PhD
Centre for mRNP Biogenesis and Metabolism, Department of Molecular Biology, Aarhus University, Aarhus, Denmark

Springer Science+Business Media, LLC

Landes Bioscience

Springer Science+Business Media, LLC
Landes Bioscience

Springer Science+Business Media, LLC, 233 Spring Street, New York, New York 10013, USA
http://www.springer.com

Please address all inquiries to the publishers:
Landes Bioscience, 1806 Rio Grande, Austin, Texas 78701, USA
Phone: 512/ 637 6050; FAX: 512/ 637 6079
http://www.landesbioscience.com

The chapters in this book are available in the Madame Curie Bioscience Database.
http://www.landesbioscience.com/curie

RNA Exosome, edited by Torben Heick Jensen. Landes Bioscience / Springer Science+Business Media, LLC dual imprint / Springer series: Advances in Experimental Medicine and Biology.

ISBN 978-1-4419-7840-0

Library of Congress Cataloging-in-Publication Data

RNA exosome / edited by Torben Heick Jensen.
p. ; cm. -- (Advances in experimental medicine and biology ; v. 702)
Includes bibliographical references and index.
ISBN 978-1-4419-7840-0 (hardback)
1. Ribonucleases. 2. RNA--Metabolism. 3. Ribosomes. I. Jensen, Torben Heick, 1965- , editor. II. Series: Advances in experimental medicine and biology ; v. 702.
[DNLM: 1. Exoribonucleases. 2. RNA--metabolism. 3. RNA Stability. W1 AD559 v.702 2012 / QU 136]
QP609.R53R63 2012
572.8'8--dc22

2010045268

PREFACE

The diversity of RNAs inside living cells is amazing. We have known of the more "classic" RNA species: mRNA, tRNA, rRNA, snRNA and snoRNA for some time now, but in a steady stream new types of molecules are being described as it is becoming clear that most of the genomic information of cells ends up in RNA. To deal with the enormous load of resulting RNA processing and degradation reactions, cells need adequate and efficient molecular machines. The RNA exosome is arising as a major facilitator to this effect. Structural and functional data gathered over the last decade have illustrated the biochemical importance of this multimeric complex and its many co-factors, revealing its enormous regulatory power. By gathering some of the most prominent researchers in the exosome field, it is the aim of this volume to introduce this fascinating protein complex as well as to give a timely and rich account of its many functions.

The exosome was discovered more than a decade ago by Phil Mitchell and David Tollervey by its ability to trim the 3'end of yeast, *S. cerevisiae*, 5.8S rRNA. In a historic account they laid out the events surrounding this identification and the subsequent birth of the research field. In the chapter by Kurt Januszyk and Christopher Lima the structural organization of eukaryotic exosomes and their evolutionary counterparts in bacteria and archaea are discussed in large part through presentation of structures. The functional implications of many of these are discussed in subsequent chapters dealing with the organizations and utilities of archea (Elena Evguenieva-Hackenberg), protist (Christine Clayton and Antonio Estevez) and plant (Heike Lange and Dominique Gagliardi) exosomes.

Exosomes gain their functional properties by associating with exo- and endo-nucleolytic subunits as well as with additional enzymes like RNA helicases and poly(A) polymerases. Collectively, this results in a flexible molecular machine capable of dealing with a multitude of cellular RNA substrates of both nuclear and cytoplasmic origin. Interestingly, prokaryotes employ a basic set of enzymes and therefore may illustrate the evolutionary origin of the eukaryotic system. These issues are discussed in three chapters: by Aleksander Chlebowski, Rafal Tomecki, Maria Eugenia Gas Lopez, Bertrand Seraphin, and Andrzej Dziembowski; by Daneen Schaeffer, Amanda Clark, Alejandra Klauer, Borislava Tsanova, and Ambro van Hoof as well as by Scott Butler and Phil

Mitchell. Recently discovered roles in the elimination of transcriptional noise and in heterochromatization underscore the tremendous flexibility of the RNA exosome. The chapters by Coy and Vasiljeva and Rougemaille and Libri deal with these effects of the exosome on gene silencing.

Finally, given the essential nature of the complex, it may come as no surprise that it is implicated with different human disease states. Even before exosome function was described in *S. cerevisiae*, sera of certain autoimmune patients identified a protein complex which later turned out to be the human exosome. The chapter by Staals and Pruijn gives a historic perspective on this parallel discovery of exosome and discusses the occurrence of autoantibodies to exosome components in autoimmune diseases and the connection of the exosome with cancer.

In closing, I wish to thank all contributing authors for doing a fantastic job of portraying our current structural and mechanistic knowledge about the RNA exosome.

Torben Heick Jensen, PhD
Centre for mRNP Biogenesis and Metabolism, Department of Molecular Biology,
Aarhus University, Aarhus, Denmark

ABOUT THE EDITOR...

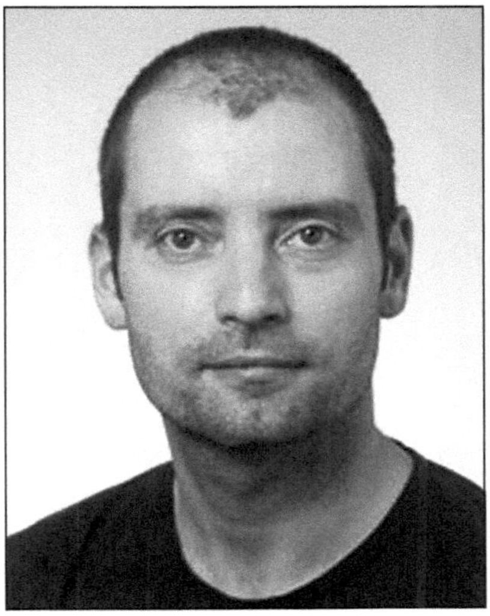

TORBEN HEICK JENSEN is a Professor at the Department of Molecular Biology at Aarhus University, Denmark. He is directing the Centre for mRNP Biogenesis and Metabolism established by the Danish National Research Foundation and employing a total of ~30 RNA researchers at all levels. Main research interests include the coupling between transcription and RNA biogenesis, quality control of RNA and the occurrence and potential function of pervasive transcription of eukaryotic genomes.

PARTICIPANTS

J. Scott Butler
Departments of Microbiology
and Immunology
Center for RNA Biology:
From Genome to Medicine
University of Rochester Medical Center
Rochester, New York
USA

Aleksander Chlebowski
Institute of Genetics and Biotechnology
Faculty of Biology
University of Warsaw
Warsaw
Poland

Amanda Clark
Department of Microbiology
and Molecular Genetics
University of Texas
Health Science Center
Houston, Texas
USA

Christine Clayton
Center for Molecular Biology
University of Heidelberg
Heidelberg
Germany

Sarah Coy
Department of Biochemistry
University of Oxford
Oxford
UK

Andrzej Dziembowski
Institute of Genetics and Biotechnology
Faculty of Biology
University of Warsaw
Warsaw
Poland

Antonio Estevez
Institute of Parasitology and Biomedicine
"Lopez-Neyra" CSIC
(Spanish National Research Council)
Armilla, Granada
Spain

Elena Evguenieva-Hackenberg
Institute of Microbiology
and Molecular Biology
University of Giessen
Giessen
Germany

Dominique Gagliardi
Institute of Plant Molecular Biology
University of Strasbourg
Strasbourg
France

María Eugenia Gas López
Institute of Genetics and Molecular
and Cellular Biology
Illkirch
France

Kurt Januszyk
Structural Biology Program
Sloan-Kettering Institute
New York, New York
USA

Torben Heick Jensen
Centre for mRNP Biogenesis
and Metabolism
Department of Molecular Biology
Aarhus University
Aarhus
Denmark

A. Alejandra Klauer
Department of Microbiology
and Molecular Genetics
University of Texas
Health Science Center
Houston, Texas
USA

Heike Lange
Institute of Plant Molecular Biology
University of Strasbourg
Strasbourg
France

Domenico Libri
LEA Laboratory of Nuclear RNA
Metabolism
Center for Molecular Genetics
Gif-sur-Yvette
France

Christopher D. Lima
Structural Biology Program
Sloan-Kettering Institute
New York, New York
USA

Phil Mitchell
Department of Molecular Biology
and Biotechnology
The University of Sheffield
Western Bank, Sheffield
UK

Ger J.M. Pruijn
Department of Biomolecular Chemistry
Radboud University
Nijmegen
The Netherlands

Mathieu Rougemaille
LEA Laboratory of Nuclear RNA
Metabolism
Center for Molecular Genetics
Gif-sur-Yvette
France

Daneen Schaeffer
Department of Microbiology
and Molecular Genetics
University of Texas
Health Science Center
Houston, Texas
USA

Bertrand Séraphin
Institute of Genetics and Molecular
and Cellular Biology
Illkirch
France

Raymond H.J. Staals
Department of Biomolecular Chemistry
Radboud University
Nijmegen
The Netherlands

David Tollervey
Wellcome Trust Centre for Cell Biology
University of Edinburgh
Edinburgh
Scotland

Rafał Tomecki
Institute of Genetics and Biotechnology
Faculty of Biology
University of Warsaw
Warsaw
Poland

Borislava Tsanova
Department of Microbiology
and Molecular Genetics
University of Texas
Health Science Center
Houston, Texas
USA

Ambro van Hoof
Department of Microbiology
and Molecular Genetics
University of Texas
Health Science Center
Houston, Texas
USA

Lidia Vasiljeva
Department of Biochemistry
University of Oxford
Oxford
UK

CONTENTS

CHAPTER 1

FINDING THE EXOSOME

Phil Mitchell*[,1] and David Tollervey*[,2]

[1]*Department of Molecular Biology and Biotechnology, University of Sheffield, Sheffield, UK;*
[2]*Wellcome Trust Centre for Cell Biology, University of Edinburgh, Edinburgh, Scotland, UK.*
**Correspondence—Emails: p.j.mitchell@sheffield.ac.uk, d.tollervey@ed.ac.uk*

Abstract: We describe the events surrounding the identification of the exosome complex and the subsequent early development of the field. Like many scientific discoveries, the initial identification and characterization of the exosome was based on a combination of skill, good fortune—and the availability of cutting edge technology.

INTRODUCTION

The early 1990s were in some respects a frustrating time for ribosome research. On the one hand, established plasmid-based rDNA systems enabled the precise mapping and mutagenesis analysis of processing sites within the pre-rRNA sequence and a steady flow of processing factors were being identified by genetic, biochemical and bioinformatic approaches. However, genetic depletion of the identified trans-acting factors typically caused a common set of defects in pre-rRNA processing that led to a general loss of rRNA levels and it was therefore impossible to identify the specific molecular function of the protein. We therefore decided to concentrate on trying to identify enzymes that participate in the pre-rRNA processing reactions. The identification of these factors promised opportunities for more insightful experiments such as in vitro biochemical analyses, targeted mutagenesis studies and studies on the regulation of the pathway.

RNA Exosome, edited by Torben Heick Jensen.

FROM NORTHERNS TO NUCLEASES

We were therefore interested in screening banks of conditional yeast mutants for those showing blocks in specific steps in the pre-rRNA processing pathway, rather than just the loss of the pre-rRNAs. The problem was that generating and testing banks of mutants required and still requires, a great deal of work. Happily, at about this time Zoe Lygerou, in the group of Bertrand Séraphin also working at EMBL, was attempting to identify mutants defective in the interaction between the U4 and U6 small nuclear RNAs. To this end they had generated a collection of around 250 temperature-sensitive (ts) lethal yeast strains, extracted RNA from each and resolved the RNA on nondenaturing gels suitable for separating low molecular weight RNAs. The gels were far from ideal for the analysis of pre-rRNA processing defects and carried RNA from only around 250 strains, not all of which had given usable separation (see Fig. 1). The blots were, however, at hand and so we screened them by northern hybridization for defects in processing of the similarly sized 5.8S rRNA.

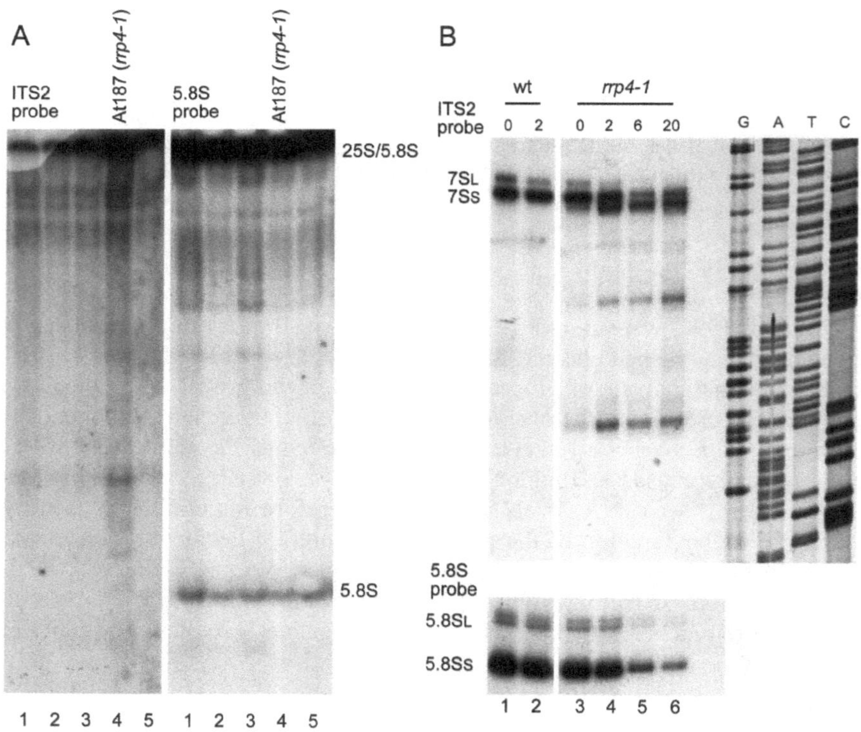

Figure 1. Identification of strain At187 harboring the *rrp4–1* mutation. A) RNA was extracted under nondenaturing conditions from a collection of temperature-sensitive mutant yeast strains, separated by native gel electrophoresis and analyzed by northern hybridization. Due to the native gel conditions 5.8S rRNA largely remains associated with the 25S rRNA, but the ITS2 probe in strain At187 lit bands above the 5.8S rRNA. B: Subsequent analyses on denaturing gels revealed that these represented the accumulation of 3′ extended species. A sequencing ladder was run on the same gel and transferred to the northern filter as a size marker. Following back-crossing, the *RRP4* gene was cloned by complementation of the ts-phenotype of the *rrp4–1* mutation that is carried by this strain.

A small number of the ts mutants showed apparent alterations in 5.8S rRNA levels and/or accumulation of potential precursors and for two of these, reanalysis of the mutant by RNA extraction and separation on denaturing gels confirmed the suggested phenotype. The first mutant to be characterized was found to be defective in 5′ maturation of both pre-tRNA and 5.8S rRNA. The corresponding gene was cloned by complementation of the ts growth phenotype. Subsequent analysis revealed that it encoded a protein that was a common component of the endoribonucleases RNase P and RNase MRP, the first such protein to be identified.[1] RNase P cleaves the 5′ end of tRNAs (reviewed in ref. 2), whereas RNase MRP cleaves the pre-rRNA at site A3 both in vivo and in vitro.[1,3,4] The protein is essential, conserved to humans and was designated Pop1 (for processing of precursor RNAs). This finding led us to propose that RNase MRP and RNase P have evolved from a common enzyme involved in ribosome synthesis.[5]

Encouraged by these results, we then turned our attention to the second ts mutation. In this case, better characterization of the pre-rRNA processing phenotype revealed a ladder of 3′-extended forms of the 5.8S rRNA. The corresponding gene was cloned and designated as *RRP4* (rRNA processing defective); like Pop1, the Rrp4 protein was found to be essential and well conserved to humans. By expressing an epitope-tagged form of Rrp4 in yeast and purifying the protein from cell lysates, we showed that Rrp4 had an associated 3′ to 5′ exoribonuclease activity. These observations led us to propose that the 3′ end of the mature 5.8S rRNA is generated by a 3′ to 5′ exonuclease acting from the downstream C2 cleavage site.[6]

A GLIMPSE OF THE EXOSOME

Glycerol gradient ultracentrifugation analyses indicated that Rrp4 is a component of a moderately sized complex with an estimated mass of approximately 300–400kDa. To identify other protein components of this complex, the epitope-tagged Rrp4 was affinity purified from peak gradient fractions. These studies predated the establishment of the TAP tagging procedure[7] and the purifications were performed using a noncleavable protein A tag (see Fig. 2). After binding to IgG beads, retained proteins were nonspecifically eluted with acetic acid and resolved by SDS-PAGE. A significant problem with early analyses was the predominance of IgG molecules on the protein gels due to acid leaching from the beads. This problem was partially minimized by using gel loading buffers lacking mercaptoethanol. At this time, the identification of proteins from excised gel bands was far from routine but, fortunately, Mathias Mann was then developing this approach and agreed to analyze the samples. Despite the large amount of IgG present in the samples, Andrej Schevchenko in Matthias Mann's lab identified four other proteins from the first gels, along with Rrp4. For consistency, we designated these proteins Rrp41, Rrp42, Rrp43 and Rrp44. The band containing Rrp41 was much stronger than expected in the initial gels, but 6 peptides sequenced by mass spectrometry (a lot for that time) all corresponded to Rrp41, leading us to conclude (incorrectly as it turned out) that it was present in multiple copies in the complex.[8] Subsequent analyses were to reveal the presence of multiple, similarly sized proteins in the complex. Reassuringly, genetic depletion of each of the identified proteins elicited the same 5.8S rRNA processing phenotype observed in the original *rrp4–1* ts mutant. To our delight, one of the proteins identified, Rrp44, was clearly homologous to bacterial RNase R, a member of the RNase II family of processive, hydrolytic exonucleases. We obtained a

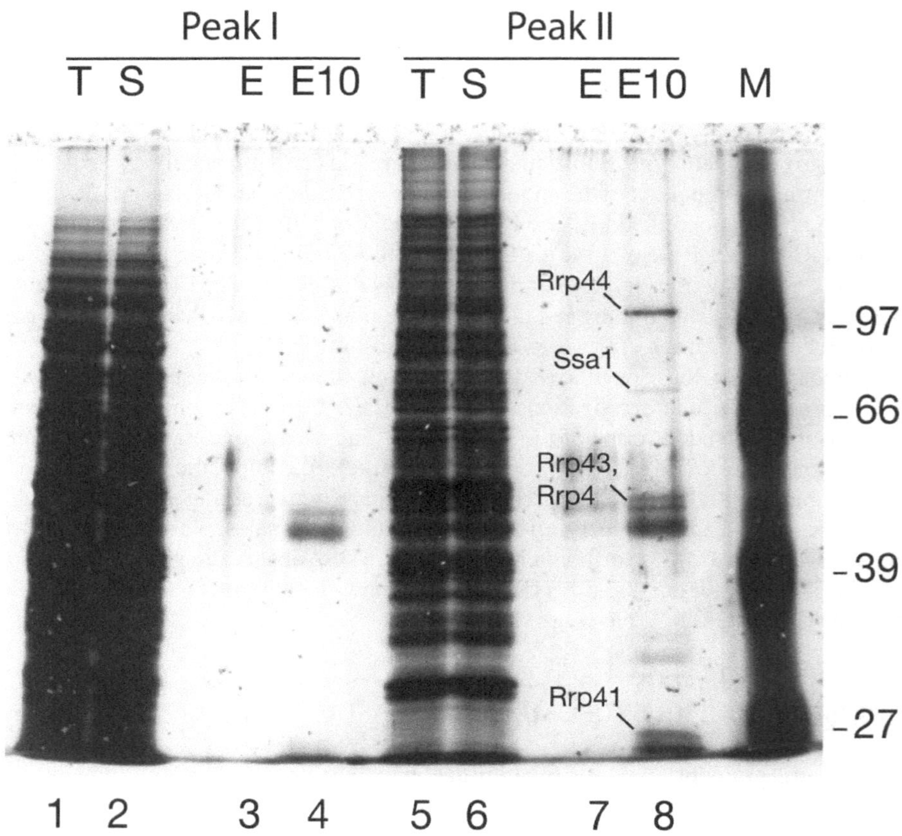

Figure 2. First analysis of the exosome complex. Analytical gel of the first exosome preparation that was purified over a glycerol gradient. Western analyses of the gradient fractions identified two peaks of Rrp4, which were then purified by affinity chromatography on IgG sepharose beads, eluted with acetic acid and the proteins resolved by SDS-PAGE. Lane 1, pooled gradient fractions from peak I. Lane 2, supernatant from peak I. Lane 3, acid eluate from peak I. Lane 4, 10-fold enriched eluate fraction. Lanes 5–8, equivalent samples from peak II. The sizes (kDa) of molecular weight markers are indicated. Protein bands in the peak II eluate fraction that were subsequently identified by mass spectrometry are labeled.

recombinant expression construct for Rrp44 from the lab of Takeharu Nishimoto, who had previously identified the *S.pombe* homologue of Rrp44 (Dis3) through interactions with the Ran GTPase. Using this, we were able to show that Rrp44 had the predicted processive, exoribonuclease activity. We now had a candidate enzyme for the observed 5.8S rRNA processing activity.

However, we noted that the processive activity of recombinant Rrp44 did not correlate with the distributive activity that was associated with Rrp4 pull-downs from yeast cell extracts, suggesting that either the activity of Rrp44 was altered upon assembly into the complex or that at least one other component of the complex also had catalytic activity. Both explanations eventually turned out to be correct. Before analyzing the Rrp4 complex, we had expressed Rrp4 in *E. coli* and tested the recombinant protein for exonuclease activity. We observed a distributive activity very similar to that seen with the affinity purified yeast complex, which convinced us that Rrp4 was a distributive exonuclease. This result proved to be irreproducible and presumably reflected the presence of a contaminant activity. Studies by Andrej Dziembowski and Bertrand Séraphin[9] later showed that the distributive in vitro activity of the exosome was probably actually due to Rrp6, which was subsequently identified as a component of the complex.

THE COMPLETE COMPLEX

Bioinformatics analyses also played an important part in the characterization of the exosome complex. Initial BLAST analyses revealed that Rrp41 was homologous to another *E. coli* nuclease, RNase PH. BLAST searches failed to identify homologues for Rrp42 and Rrp43, but Sara Mian identified a family of yeast proteins related to RNase PH including Rrp41, Rrp42 and Rrp43.[8,10] This family was also predicted to include the uncharacterized products of the *YDR280w* and *YGR095c* genes. Around this time Roy Parker and Reed Wickner discovered that Rrp41 was identical to the product of the previously identified but uncloned *SKI6* gene, which the Parker lab then showed to function in cytoplasmic mRNA turnover.[11,12] They also realized that yet another yeast protein, Mtr3, was homologous to Rrp41/Ski6 and RNase PH.

Based on this sequence homology, Christine Allmang made conditional mutants of the two uncharacterized genes and showed that depletion of the encoded proteins also led to the accumulation of 3′ extended forms of 5.8S rRNA.[13] We named these genes *RRP45* and *RRP46*, respectively. Analysis of the *mtr3–1* ts mutant, obtained from the lab of Alan Tartakoff, revealed a similar defect in 5.8S rRNA maturation. Exosome complexes from Archeae and plants exhibit exonuclease activity mediated by RNase PH related subunits,[14-16] but this does not appear to be the case for other eukaryotic systems tested.

In the meantime, we had made two further modifications to the exosome purification procedures. Firstly, we resolved the cell lysates by ion exchange chromatography prior to SDS-PAGE analysis. This enabled a significant scale-up of the experiments and provided a purification step that was considerably more rapid and as equally effective as resolving lysates through glycerol gradients. We subsequently adopted the technique, developed by Dirk Görlich, for eluting proteins retained on IgG beads with a gradient of increasing $MgCl_2$ concentration.[17] Together, these two approaches produced gels that allowed the identification of Rrp45, Rrp46, Mtr3 and two newly identified proteins, Rrp40 and Csl4, in addition to Rrp4 and Rrp41–44. Genetic depletion of Rrp40 and Csl4 also led to the same 5.8S rRNA processing phenotype. The gels of the gradient eluate fractions also revealed that Rrp44 was not as stably bound to the complex as the other identified components, demonstrating the existence of a stable nonomeric "core". We were also able to biochemically resolve two distinct forms of the complex that differed by the presence or absence of the nuclear protein Rrp6, another exoribonuclease that had been shown to function in 5.8S rRNA processing by the group of Scott Butler.[18] Subsequently, we and others identified three additional substoichiometric components of the complex, the cytoplasmic GTPase Ski7 and the nuclear RNA binding proteins Rrp47/Lrp1 and Mpp6.[19-24]

A BROADER PERSPECTIVE

We generated cDNA expression constructs of the gene encoding the human homologue of Rrp4 from a HeLa cDNA library that was provided by Karsten Weis, who was then in the lab of Angus Lamond at EMBL and were able to show that overexpression of the encoded protein complemented the ts growth phenotype of the *rrp4–1* mutant. Coimmunoprecipitation studies subsequently showed that a complex closely related to the yeast exosome was present in human cells, where it had initially been identified as the PM-Scl complex, a target of autoimmune antibodies in patients suffering from the unfortunate sounding polymyositis-scleroderma overlap syndrome (PM-Scl—now termed Scleromyositis).[13,25] The human exosome complex was then purified by expressing epitope-tagged hRrp4 and shown to act in ARE-mediated mRNA decay.[26]

The barrel shape of the exosome was first predicted by Ambro van Hoof and Roy Parker,[27] who insightfully suggested that the structure might be analogous to the proteasome. Support for this model was provided by structural analysis of the Bacterial PNPase, which was shown to have a domain composition strikingly similar to that of the exosome.[28,29] Finally, the crystal structures of the archaeal and then the human exosome confirmed the exosome structure as a barrel with a central cavity through which substrates pass.[16,30-32]

The distributive activity of yeast exosome complexes observed in vitro did not correlate well with genetic and biochemical data that indicated potent activity in vivo on highly structured RNA-protein complexes. The obvious hypothesis was that the exosome was largely dependent upon additional cofactors for substrate identification and activation in vivo and numerous cofactors have since been identified; reviewed in ref. 33. Recent studies have also revealed an endoribonuclease activity of Rrp44.[34-36]

Although the exosome was initially characterized as a pre-rRNA processing complex, it was immediately clear from the growth characteristics of the ts *rrp4–1* mutant that it had additional functions that are essential for normal cell growth. In addition to the studies linking the exosome to cytoplasmic mRNA turnover,[12] work in our lab and others soon revealed functions in the processing of small stable RNAs[37] and subsequently showed it to function in surveillance pathways that eliminate aberrantly processed or assembled transcripts.[38-41] Loss of these functions would nevertheless not be expected to trigger the rapid growth inhibition that can be seen in conditional exosome mutants. This may now have been largely explained by recent analyses of the plethora of cryptic noncoding RNAs; see for example reference 42 and reviews by references 33 and 43. These have revealed a fundamental function of the exosome and its cofactors as key gatekeepers of the transcriptome.

CONCLUSION

Despite many publications, the identification of a vast number of substrates and the characterization of numerous cofactors, key questions concerning the exosome remain unanswered. In particular, how are the distinct catalytic activities of the complex coordinated and how do the cofactors modulate the activity of the exosome complex at the molecular level? Resolving these questions will require a range of approaches—and should continue to provide a fruitful field of study for some time to come.

ACKNOWLEDGEMENTS

D.T. was supported by the Wellcome Trust. P.M. was supported by a research grant from the Wellcome Trust (088306/Z/09/Z).

REFERENCES

1. Lygerou Z, Mitchell P, Petfalski E et al. The POP1 gene encodes a protein component common to the RNase MRP and RNase P ribonucleoproteins. Genes Dev 1994; 8:1423-1433.
2. Evans D, Marquez SM, Pace NR. RNase P: interface of the RNA and protein worlds. Trends Biochem Sci 2006; 31:333-341.
3. Henry Y, Wood H, Morrissey JP et al. The 5′ end of yeast 5.8S rRNA is generated by exonucleases from an upstream cleavage site. EMBO J 1994; 13:2452-2463.
4. Lygerou Z, Allmang C, Tollervey D et al. Accurate processing of a eukaryotic prerRNA by RNase MRP in vitro. Science 1996; 272:268-270.
5. Morrissey JP, Tollervey D. Birth of the snoRNPs—the evolution of RNase MRP and the eukaryotic prerRNA processing system. Trends Biochem Sci 1995; 20:78-82.
6. Mitchell P, Petfalski E, Tollervey D. The 3′-end of yeast 5.8S rRNA is generated by an exonuclease processing mechanism. Genes Dev 1996; 10:502-513.
7. Rigaut G, Shevchenko A, Rutz B et al. A generic protein purification method for protein complex characterization and proteome exploration. Nat Biotechnol 1999; 17:1030-1032.
8. Mitchell P, Petfalski E, Shevchenko A et al. The exosome; a conserved eukaryotic RNA processing complex containing multiple 3′->5′ exoribonuclease activities. Cell 1997; 91:457-466.
9. Dziembowski A, Lorentzen E, Conti E et al. A single subunit, Dis3, is essentially responsible for yeast exosome core activity. Nat Struct Mol Biol 2007; 14:15-22.
10. Mian IS. Comparative sequence analysis of ribonucleases HII, III, II PH and D. Nucleic Acids Res 1997; 25:3187-3195.
11. Benard L, Carroll K, Valle RC et al. Ski6p is a homolog of RNA-processing enzymes that affects translation of nonpoly(A) mRNAs and 60S ribosomal subunit biogenesis. Mol Cell Biol 1998; 18:2688-2696.
12. Anderson JSJ, Parker RP. The 3′ to 5′ degradation of yeast mRNAs is a general mechanism for mRNA turnover that requires the SKI2 DEVH box protein and 3′ to 5′ exonucleases of the exosome complex. EMBO J 1998; 17:1497-1506.
13. Allmang C, Petfalski E, Podtelejnikov A et al. The yeast exosome and human PM-Scl are related complexes of 3′->5′ exonucleases. Genes Dev 1999; 13:2148-2158
14. Chekanova JA, Shaw RJ, Wills MA et al. Poly(A) tail-dependent exonuclease AtRrp41p from Arabidopsis thaliana rescues 5.8 S rRNA processing and mRNA decay defects of the yeast ski6 mutant and is found in an exosome-sized complex in plant and yeast cells. J Biol Chem 2000; 275:33158-66.
15. Chekanova JA, Dutko JA, Mian IS et al. Arabidopsis thaliana exosome subunit AtRrp4p is a hydrolytic 3′—>5′ exonuclease containing S1 and KH RNA-binding domains. Nucleic Acids Res 2002; 30:695-700.
16. Lorentzen E, Walter P, Fribourg S et al. The archaeal exosome core is a hexameric ring structure with three catalytic subunits. Nat Struct Mol Biol 2005; 12:575-581.
17. Görlich D, Kraft R, Kostka S et al. Importin provides a link between nuclear protein import and U snRNA export. Cell 1996; 87:21-32.
18. Briggs MW, Burkard KT, Butler JS. Rrp6p, the yeast homologue of the human PM-Scl 100-kDa autoantigen, is essential for efficient 5.8 S rRNA 3′ end formation. J Biol Chem 1998; 273:13255-13263.
19. Mitchell P, Petfalski E, Houalla R et al. Rrp47p is an exosome-associated protein required for the 3′ processing of stable RNAs. Mol Cell Biol 2003; 23:6982-6992.
20. Peng WT, Robinson MD, Mnaimneh S et al. A panoramic view of yeast noncoding RNA processing. Cell 2003; 113:919-33.
21. Araki Y, Takahashi S, Kobayashi T et al. Ski7p G protein interacts with the exosome and the Ski complex for 3′-to-5′ mRNA decay in yeast. EMBO J 2001; 20:4684-4693.
22. van Hoof A, Staples RR, Baker RE et al. Function of the ski4p (Csl4p) and Ski7p proteins in 3′-to-5′ degradation of mRNA. Mol Cell Biol 2000; 20:8230-8243.
23. Milligan L, Decourty L, Saveanu C et al. A yeast exosome cofactor, Mpp6, functions in RNA surveillance and in the degradation of noncoding RNA transcripts. Mol Cell Biol 2008; 28:5446-57.
24. Schilders G, Raijmakers R, Raats JM et al. MPP6 is an exosome-associated RNA-binding protein involved in 5.8S rRNA maturation. Nucleic Acids Res 2005; 33:6795-6804.

25. Brouwer R, Pruijn GJ, van Venrooij WJ. The human exosome: an autoantigenic complex of exoribonucleases in myositis and scleroderma. Arthritis Res 2001; 3:102-106.
26. Chen CY, Gherzi R, Ong SE et al. AU binding proteins recruit the exosome to degrade ARE-containing mRNAs. Cell 2001; 107:451-464.
27. van Hoof A, Parker R. The exosome: a proteasome for RNA? Cell 1999; 99:347-350.
28. Symmons MF, Jones GH, Luisi BF. A duplicated fold is the structural basis for polynucleotide phosphorylase catalytic activity, processivity and regulation. Structure Fold Des 2000; 8:1215-26.
29. Symmons MF, Williams MG, Luisi BF et al. Running rings around RNA: a superfamily of phosphate-dependent RNases. Trends Biochem Sci 2002; 27:11-18.
30. Lorentzen E, Conti E. Structural basis of 3′ end RNA recognition and exoribonucleolytic cleavage by an exosome RNase PH core. Mol Cell 2005; 20:473-81.
31. Liu Q, Greimann JC, Lima CD. Reconstitution, activities and structure of the eukaryotic RNA exosome. Cell 2006; 127:1223-1237.
32. Bonneau F, Basquin J, Ebert J et al. The yeast exosome functions as a macromolecular cage to channel RNA substrates for degradation. Cell 2009; 139:547-559.
33. Houseley J, Tollervey D. The many pathways of RNA degradation. Cell 2009; 136:763-776.
34. Lebreton A, Tomecki R, Dziembowski A et al. Endonucleolytic RNA cleavage by a eukaryotic exosome. Nature 2008; 456:993-996.
35. Schneider C, Leung E, Brown J et al. The N-terminal PIN domain of the exosome subunit Rrp44 harbors endonuclease activity and tethers Rrp44 to the yeast core exosome. Nucleic Acids Res 2009; 37:1127-1140.
36. Schaeffer D, Tsanova B, Barbas A et al. The exosome contains domains with specific endoribonuclease, exoribonuclease and cytoplasmic mRNA decay activities. Nat Struct Mol Biol 2009; 16:56-62.
37. Allmang C, Kufel J, Chanfreau G et al. Functions of the exosome in rRNA, snoRNA and snRNA synthesis. EMBO J 1999; 18:5399-5410.
38. Allmang C, Mitchell P, Petfalski E et al. Degradation of ribosomal RNA precursors by the exosome. Nucleic Acids Res 2000; 28:1684-1691.
39. van Hoof A, Frischmeyer PA, Dietz HC et al. Exosome-mediated recognition and degradation of mRNAs lacking a termination codon. Science 2002; 295:2262-2264.
40. Hilleren P, McCarthy T, Rosbash M et al. Quality control of mRNA 3′-end processing is linked to the nuclear exosome. Nature 2001; 413:538-542.
41. Bousquet-Antonelli C, Presutti C, Tollervey D. Identification of a regulated pathway for nuclear premRNA turnover. Cell 2000; 102:765-775.
42. Wyers F, Rougemaille M, Badis G et al. Cryptic Pol II transcripts are degraded by a nuclear quality control pathway involving a new poly(A) polymerase. Cell 2005; 121:725-737.
43. Belostotsky D. Exosome complex and pervasive transcription in eukaryotic genomes. Curr Opin Cell Biol 2009; 21:352-358.

CHAPTER 2

STRUCTURAL COMPONENTS AND ARCHITECTURES OF RNA EXOSOMES

Kurt Januszyk and Christopher D. Lima*

Structural Biology Program, Sloan-Kettering Institute, New York, New York, USA.

**Corresponding Author: Christopher D. Lima—Email: limac@mskcc.org*

Abstract: A large body of structural work conducted over the past ten years has elucidated mechanistic details related to 3′ to 5′ processing and decay of RNA substrates by the RNA exosome. This chapter will focus on the structural organization of eukaryotic exosomes and their evolutionary cousins in bacteria and archaea with an emphasis on mechanistic details related to substrate recognition and to 3′ to 5′ phosphorolytic exoribonucleolytic activities of bacterial and archaeal exosomes as well as the hydrolytic exoribonucleolytic and endoribonucleolytic activities of eukaryotic exosomes. These points will be addressed in large part through presentation of crystal structures of phosphorolytic enzymes such as bacterial RNase PH, PNPase and archaeal exosomes and crystal structures of the eukaryotic exosome and exosome sub-complexes in addition to standalone structures of proteins that catalyze activities associated with the eukaryotic RNA exosome, namely Rrp44, Rrp6 and their bacterial counterparts.

INTRODUCTION

Enzymes that catalyze 3′ to 5′ RNA decay share evolutionary relationships throughout prokaryotic, archaeal and eukaryotic phylogeny (Fig. 1). 3′ to 5′ RNA decay is promoted by three distinct classes of enzymes that catalyze exoribonuclease activity in bacteria. One includes two related enzymes, RNase II and RNase R, which catalyze processive hydrolytic RNA decay. Another class includes the enzyme RNase D which catalyzes distributive hydrolytic RNA decay. The third class includes PNPase, a processive phosphorolytic exoribonuclease that is associated with the degradosome, a RNA decay complex comprised of PNPase, the endoribonuclease RNase E, the RNA helicase RhlB and enolase.[1-2] PNPase is a multi-domain protein that homooligomerizes

RNA Exosome, edited by Torben Heick Jensen.

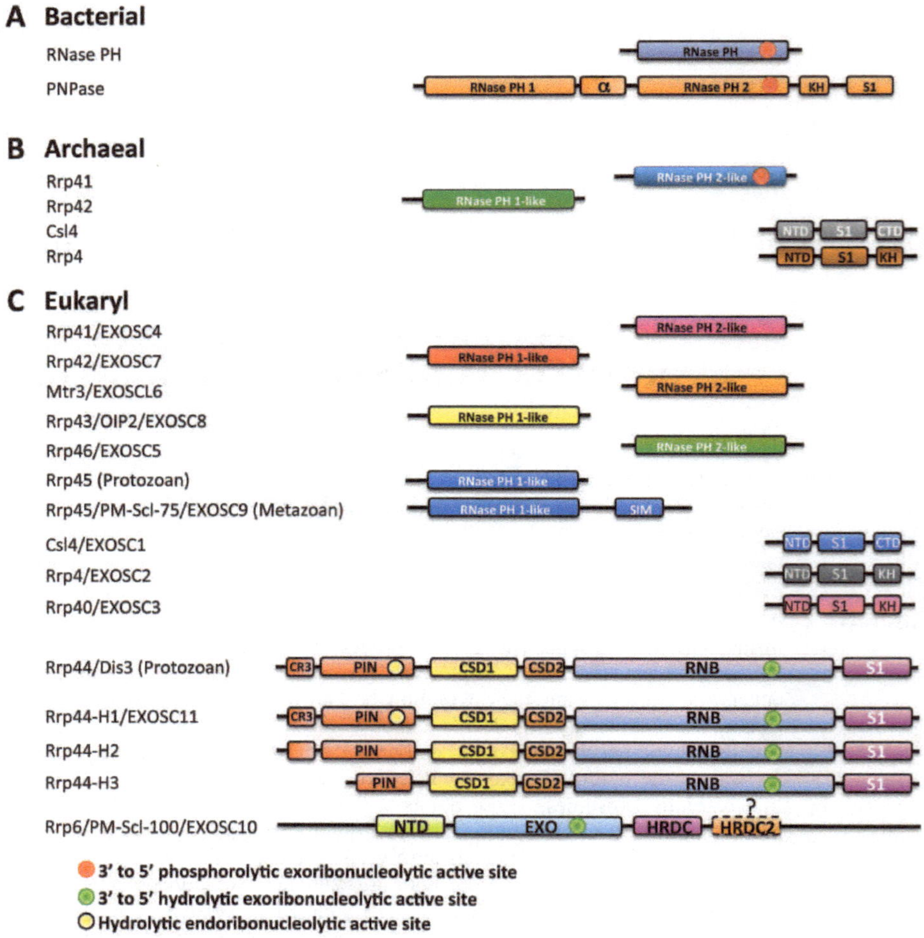

Figure 1. Schematics of domains in 'exosomes' from bacteria, archaea and eukaryotes. A) Bacterial RNase PH and PNPase subunits. Bacterial RNase PH contains a 3′ to 5′ phosphorolytic exoribonucleolytic active site (red circle). Bacterial PNPase contains five domains: RNase PH 1, the alpha domain, RNase PH 2, a KH domain and an S1 domain. A 3′ to 5′ phosphorolytic exoribonucleolytic active site resides in the RNase PH 2 domain (red circle). B) Archaeal exosome subunits. Archaeal exosomes include four subunits: Rrp41, that contains a RNase PH 2-like domain with a 3′ to 5′ phosphorolytic exoribonucleolytic active site (red circle), Rrp42, that contains a RNase PH 1-like domain and either Csl4 or Rrp4. Csl4 contains an N-Terminal Domain (NTD), a S1 domain and a KH domain. Rrp4 contains an N-Terminal Domain (NTD), a S1 domain and a Carboxy Terminal Domain (CTD). C) Eukaryotic exosome subunits. Protozoan and metazoan exosomes contain either ten or eleven components consisting of nine catalytically inert core components (Rrp41, Rrp42, Mtr3, Rrp43, Rrp46, Rrp45, Csl4, Rrp4 and Rrp40) and two active components Rrp44 and Rrp6. Alternative names for each of the eukaryotic subunits are included. Rrp44 contains five annotated domains: a PIN (PIlus N terminal) domain with a Cysteine-Rich sequence (CR3), two Cold Shock Domains (CSD1 and CSD2), a Ribo Nuclease Binding (RNB) domain and an S1 domain. The "hydrolytic" endoribonucleolytic activity is located within the PIN domain (yellow circle) and the processive 3′ to 5′ hydrolytic exoribonucleolytic activity resides in the RNB domain (green circle). Metazoan exosomes may vary in their utilization of Rrp44 (e.g., three in human H1, H2 and H3). Rrp6 contains three known domains: NTD (N-Terminal Domain), EXO (EXOribonuclease domain) and HRDC (Homology to RNase D domain C-terminal). It is hypothesized that a second HRDC domain (HRDC2) may be located C-terminal to HRDC. 3′ to 5′ distributive hydrolytic endoribonucleolytic activity is located within the EXO domain (green circle).

to form a ring-like structure with a central channel that harbors the phosphorolytic active sites.

Archaeal exosomes are processive phosphorolytic enzymes that share mechanistic and structural similarities to bacterial PNPase[3-4] (Fig. 1B). Archaeal exosomes are composed of up to four individually encoded proteins that oligomerize to form an analogous structure to PNPase, although in this instance intact exosomes form by oligomerization of six subunits that make the ring and three additional subunits that cap the ring.[5-6] As with PNPase, archaeal exosomes have a central channel through which the RNA substrate must pass to gain access to the phosphorolytic active sites.[7-8]

The eukaryotic exosome core is architecturally similar to PNPase and archaeal exosomes, although it is more complex because it is composed of nine individually encoded subunits.[9] The eukaryotic exosome also differs fundamentally from PNPase and archaeal exosomes, because it is not a phosphorolytic enzyme and instead has developed the ability to directly associate with Rrp44 and Rrp6, hydrolytic exoribonucleases that share evolutionary relationships to bacterial RNase II/R and RNase D, respectively.[9-11]

In this chapter, we will describe the individual domains and overall architectures of enzymes and proteins that contribute to 3′ to 5′ decay through formation of exosomes or exosome-related complexes in bacterial, archaeal and eukaryl organisms with an emphasis on what is currently known about their respective catalytic mechanisms and how the architecture of the intact exosome cores impacts their activities and function.

GLOBAL STRUCTURE OF THE EXOSOME

RNase PH, PNPase, archaeal and eukaryotic exosome cores are composed of evolutionarily related domains (Fig. 1) that oligomerize to form rings with central pores large enough to accommodate single stranded RNA (Fig. 2). RNase PH achieves this architecture through oligomerization of six RNase PH proteins (Fig. 1), resulting in a pseudo-hexameric ring with three-fold symmetry (Fig. 2A).[12-14] The RNase PH ring includes six phosphorolytic active sites that are located in the interface between respective RNase PH proteins. The head to tail arrangement of RNase PH proteins around the ring generates a molecular two fold axis that situates three active sites on the bottom of the ring, another three active sites on the top of the ring and RNA binding surfaces situated within the pore (Fig. 3A).

Similar to RNase PH, PNPase forms a related pseudo-hexameric ring through oligomerization of three PNPase molecules that contain an N-terminal RNase PH-like domain which we term RNase PH 1, an alpha domain, a second RNase PH-like domain which we term RNase PH 2 which is then followed by a KH domain and an S1 domain (Fig. 1A). The RNase PH 2 domain contains residues responsible for phosphorolytic activity, while the amino terminal RNase PH 1 domain is catalytically inactive.[2] The phosphorolytic active site and RNA binding surfaces are formed at the interface between the RNase PH 2 and RNase PH 1 domains (Fig. 3B). Because only one RNA PH-like domain in PNPase contains residues that form the phosphorolytic active site, only three active sites are formed in PNPase. In addition, the RNase PH 2 domain of PNPase is partially occluded from solvent by the alpha domain (bottom orientation) while additional putative RNA binding surfaces are formed by the KH and S1 domains (top orientation, Figs. 2B and 3B).

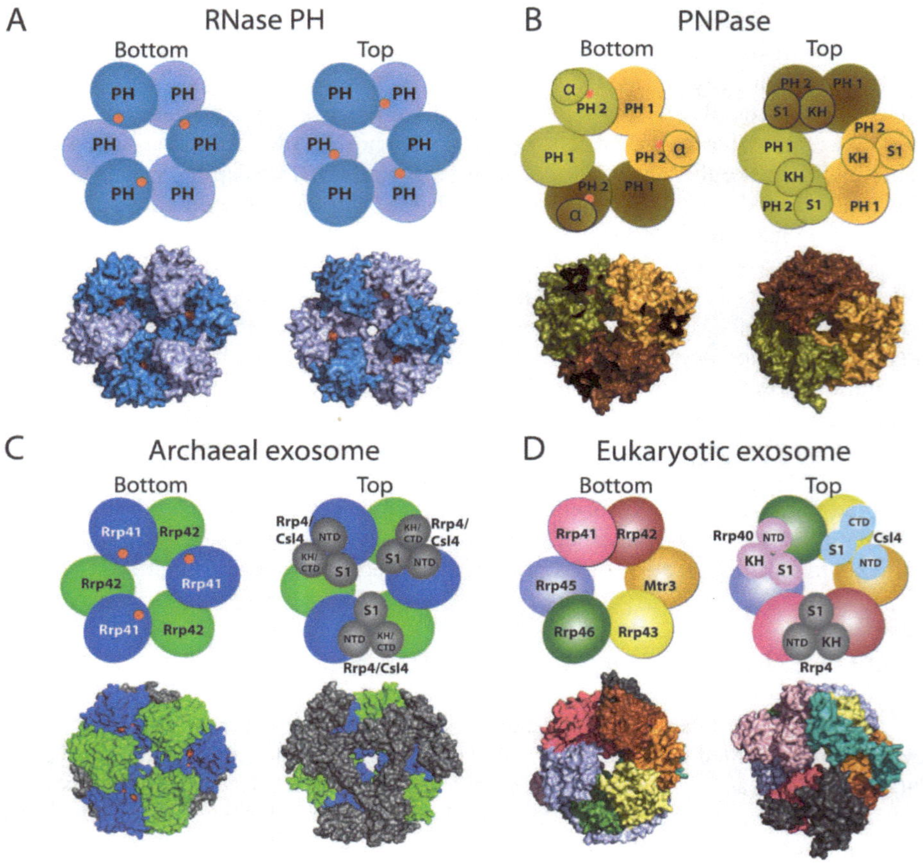

Figure 2. 'Exosomes' from bacteria, archaea and eukaryotes have a similar architecture. RNase PH, PNPase, archaeal exosome and eukaryotic core exosome structures and schematics are depicted in two orientations which we denote bottom and top. Architectures emphasize a six-component ring with or without phosphorolytic active sites (shown as red dots in cartoon representation and red surfaces in the surface representations of the respective structures). A) RNase PH. The *Aquifex aeolicus* RNase PH structure (PDB ID = 1UDN) forms a homohexamer of PH subunits (colored dark blue and light blue, for clarity). B) PNPase. The *S. antibioticus* PNPase structure (PDB ID = 1E3P) forms a homotrimer. PNPase protomers are colored light yellow, dark yellow and light brown to distinguish the homotrimer of RNase PH 1-like (PH 1) and RNase PH 2-like (PH 2) domains. The α domain was omitted to enable visualization of the phosphorolytic active sites in the bottom view (red dots). C) Archaeal exosome. The *S. solfataricus* archaeal exosome (PDB ID = 2JE6) is depicted with Rrp41 subunits (blue) and Rrp42 subunits (green) which form the six-component ring. Top view in the schematic shows the orientation of Csl4 (N-Terminal Domain, S1 domain and C-Terminal Domain) and Rrp4 (N-Terminal Domain, S1 domain and KH domain) labeled and shown in grey. The surface representation of the structure depicts the Rrp4-bound archaeal exosome. Residues in the active site are partially occluded from view such that the active site appears as two discontinuous red surfaces. D) Eukaryotic core exosome. The eukaryotic exosome is shown from *H. sapiens* (PDB ID = 2NN6). Subunits Rrp41 (magenta), Rrp42 (red), Mtr3 (orange), Rrp43 (yellow), Rrp46 (green), Rrp45 (blue) form the six-component ring. Subunits Rrp40 (light pink), Csl4 (cyan) and Rrp4 (grey) form the three-component cap. No phosphorolytic active site exists in the eukaryotic core exosome.

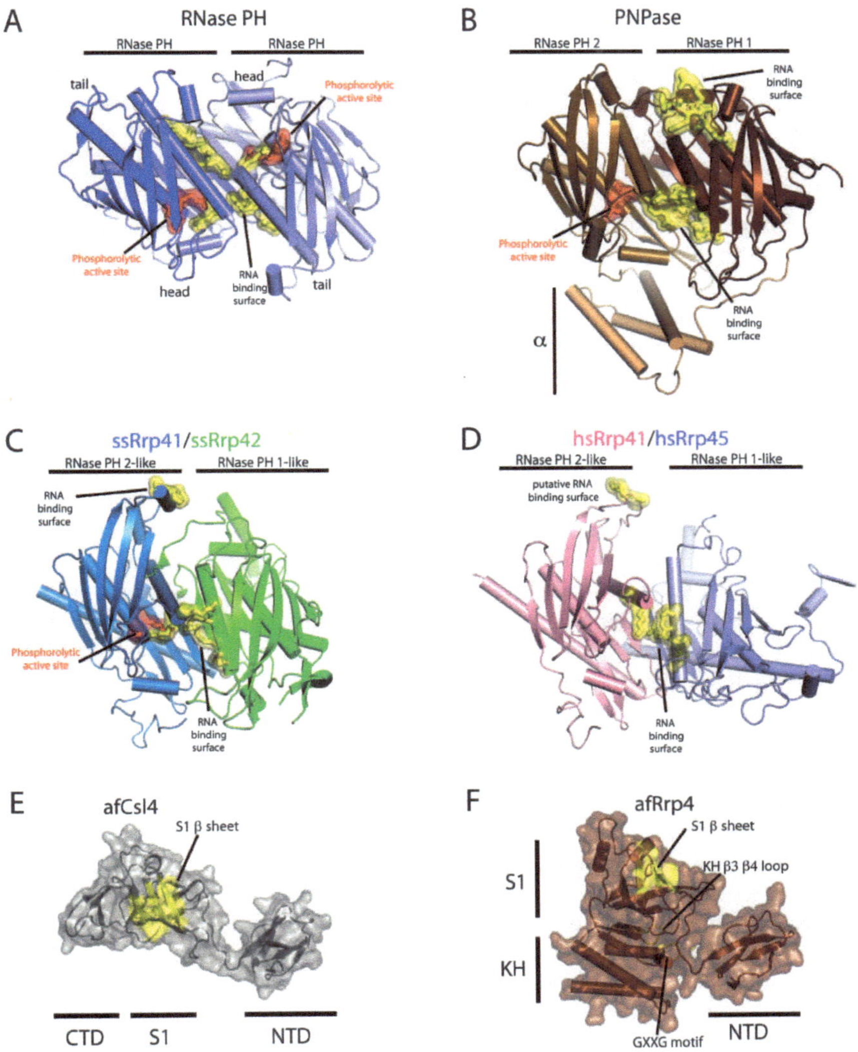

Figure 3. Structures of exosome domains. Residues in red indicate phosphate binding regions and residues in yellow highlight RNA binding surfaces. Structures depicted in cartoons with helices as tubes and β-strands as arrows. A) RNase PH homodimerization interface (PDB ID = 1UDN). Phosphate binding residues include T125 and R126. RNA binding residues: R86, R92, R96 and R99. B) PNPase RNase PH 1/RNase PH 2 domain binding interface (PDB ID = 1E3P). Phosphate binding residues include: T462 and S463. RNA binding interface residues: R100, R104, R107, R422 and R423. A second RNA binding site includes residues F84, F85, R86 and R87. C) Archaeal S. *solfataricus* exosome Rrp41/Rrp42 heterodimer interface (PDB ID = 2JE6). Phosphate binding residues are from ssRrp41: S138 and R139. RNA binding interface residues from ssRrp41 are R98 and R99 and R112, R116 and R119 (ssRrp42). The second RNA binding region includes residues R67 and H68 (from ssRrp41) D) Eukaryotic *H. sapiens* Rrp41/Rrp45 hetero-dimerization interface. Putative RNA binding interface residues include: K94, S95, R104, R108 and R111 and the second putative RNA binding region includes residues R61 and A62. Structures for the archaeal three-component cap subunits (PDB ID = 2BAO and 2BA1): E) *A. fulgidus* Csl4 and F) *A. fulgidus* Rrp4. Putative RNA binding surfaces are highlighted in yellow for the S1 domain and KH domain on a transparent surface representation. Similar structures exist for the human three-component cap subunits Csl4, Rrp4 and Rrp40 as discussed in the text (PDB ID = 2NN6).

Crystal structures from the hyper-thermophiles *Sulfolobus solfataricus*, *Archaeoglobus fulgidus* and *Pyrococcus abyssi* revealed that archaeal exosomes are composed of trimers of Rrp41-Rrp42 heterodimers which oligomerize to form pseudo-hexameric rings (Fig. 2C). Rrp41 contains residues that comprise the phosphorolytic active site that share sequence similarity with both the RNase PH 2 domain of PNPase and RNase PH (Fig. 1). Rrp42 shares sequence similarity with the RNase PH 1 domain and is catalytically inert (Figs. 2C and 3C). Analogous to bacterial PNPase, the RNA binding surfaces and active sites are located in a composite surface formed between the Rrp41 and Rrp42 heterodimer (Fig. 3C). The six-subunit rings are capped by three copies of Rrp4, Csl4, or combinations therein.[5-7,15] Rrp4 and Csl4 both contain putative sites for RNA interaction via their S1 and KH domains or S1 domain, respectively. The phosphorolytic active sites are exposed to solvent and visible at the bottom of the ring while Rrp4 or Csl4 cap the top of the ring to presumably restrict access or guide substrates into the pore for degradation (Fig. 2C, top view).

The human exosome core features a pseudo-hexameric six-component ring, three-component cap and a central pore, an architecture common to bacterial PNPase and archaeal exosomes (Figs. 1 and 2D).[9] With that said, the human exosome architecture differs somewhat from archaeal exosomes because the nine-subunit core is formed through oligomerization of nine individually encoded subunits that form the ring (Rrp41, Rrp45, Rrp42, Rrp43, Mtr3 and Rrp46) or the three-component cap (Rrp4, Rrp40 and Csl4). While it is likely that the general architecture observed for the human exosome core is predictive of other eukaryotic exosomes, subtle distinctions between protozoa and metazoa are expected; for instance, metazoan Rrp45 subunits include a large (~150 amino acid) C-terminal extension that is absent in lower eukaryotes (Fig. 1C). Interestingly in human Rrp45, this extension contains a phosphorylation-dependent SUMO interaction motif suggesting that this region of Rrp45 may be important for regulation of exosome activities or assembly.[16]

Subunits that comprise the six-component ring of eukaryotic exosomes share higher sequence and structural similarities to either archaeal Rrp41 or PNPase RNase PH 2-like proteins (Rrp41, Mtr3 and Rrp46) or archaeal Rrp42 or PNPase RNase PH 1-like proteins (Rrp42, Rrp43 and Rrp45). The six-component ring is formed by oligomerization of three distinct RNase PH 2-like and RNase PH 1-like heterodimers: Rrp41-Rrp45, Rrp43-Rrp46 and Mtr3-Rrp42. However, unlike the archaeal exosome or PNPase, both classes of eukaryotic RNase PH-like domains are devoid of catalytic activity and do not contain key catalytic residues that are conserved in PNPase or archaeal exosome phosphorolytic active sites (Figs. 1C and 2D).[5,9-11,17]

Csl4, Rrp4 and Rrp40 contain N-terminal Domains (NTD) and putative RNA binding S1 domains, but they differ with respect to inclusion of either a KH domain (as observed in Rrp4 and Rrp40) or a C-Terminal Domain (CTD), as observed for Csl4 (Fig. 1C). In addition, subunits of the cap are required to stabilize interactions between the different RNase PH-like heterodimers. Specifically in the human exosome, Rrp4 bridges interactions between Rrp41 and Rrp42, Rrp40 bridges the Rrp45 and Rrp46 interface and Csl4 interacts with Mtr3 and to a lesser extent with Rrp43 (Fig. 2D). This phenomenon of the three-component cap stabilizing the hexameric core is unique to eukaryotes, insofar as the archaeal exosome forms stable six-component rings in the absence of the three-component cap.[5] A feature unique to eukaryotic and archaeal exosomes is that the S1 domains of the three-component cap face the central pore surface, while in bacterial PNPase the KH domains face the central pore. The significance of the orientations for the S1 and KH putative RNA binding domains with respect to the central pore has not been ascertained.

RNase PH-LIKE DOMAINS IN BACTERIAL, ARCHAEAL AND EUKARYOTIC CORE EXOSOMES

RNase PH domains are comprised of a βαβα fold and are conserved in RNase PH, PNPase, archaeal exosomes and eukaryotic exosomes.[2,5-6,9] In RNase PH, two PH domains form a head to tail dimer generating a composite surface that includes residues that constitute RNA surfaces and the phosphorolytic active site (Fig. 3A). The location of the active site was determined by structures in which a sulfate ion or phosphate ion was observed in complex with RNase PH of *B. subtillis* or *A. aeolicus*, respectively.[13-14] Because RNase PH is a homodimer, two equivalent RNA binding surfaces are generated along the interdomain surface, one at the entrance of the central pore and one proximal to the active site (Fig. 3A). The functional significance of this symmetry is not understood.

In PNPase, two RNase PH-like domains are fused in a single polypeptide, but they come together in a pseudo-dimeric head to tail interaction to form a similar 'dimerization' interface to that observed in RNase PH between its respective RNase PH 1 and RNase PH 2 domains (Fig. 3B). The phosphorolytic active site is encompassed by residues from the RNase PH 2 domain and is positioned along the bottom of the inter-domain interface. Two distinct RNA binding surfaces are also present in this interface, one composed of residues from the RNase PH 1 domain at the entrance to the central pore and one proximal to the active site that is primarily composed by residues from the RNase PH 1 and RNase PH 2 domains (Fig. 3B).

The archaeal exosome is structurally analogous to PNPase with respect to the location of the two RNA binding surfaces and the phosphorolytic active site, however in this instance the interface is formed by two separately encoded subunits, archaeal Rrp41 and Rrp42 (Fig. 3C).[12-14] Rrp41 contains key catalytic residues that constitute the active site, but it also contributes residues in combination with those from Rrp42 to form one of the two RNA binding surfaces.[7,17-18] In contrast to PNPase which uses a RNase PH 1 domain surface to interact with RNA at the top of the interface, a second distinct RNA binding surface is present at the top of the heterodimeric interface in archaeal exosomes and is comprised solely by residues from its RNase PH 2-like domain, archaeal Rrp41 (Fig. 3C).

The eukaryotic exosome contains three heterodimeric RNase PH-like pairs (Rrp41-Rrp45, Rrp43-Rrp46 and Mtr3-Rrp42) that are arranged in similar head to tail configurations as observed in RNase PH, PNPase and archaeal exosomes.[9] While the key catalytic residues in RNase PH, PNPase and the archaeal exosomes are not conserved in any of the human or budding yeast RNase PH-like proteins, a few of the subunits, namely Rrp41 and Rrp45, include several basic residues that are conserved across evolution and believed to be important for RNA interactions located near the top of the Rrp41-Rrp45 heterodimeric interface and proximal to the location where the phosphorolytic active site resides in archaeal exosomes and PNPase (Fig. 3D).[9,19]

S1 AND KH CONTAINING DOMAINS IN BACTERIAL, ARCHAEAL AND EUKARYOTIC CORE EXOSOMES

Bacterial PNPase, archaeal exosomes and eukaryotic exosomes include putative RNA binding domains, KH Type I and S1, in the three-component cap subunits in their respective core complexes. KH Type I domains feature a β1-α1-α2-β2-β3-α3 secondary structure topology and a tertiary structure that consists of three beta-strands that form a sheet which packs against three alpha helices.[20] Single stranded RNA typically binds a KH Type I domain via surfaces formed by residues within helix α1, a conserved GXXG motif between helices α1 and α2, helix α2, the variable loop between strands β2 and β3 and residues within strand β2 (Fig. 3F). The S1 domain originally observed in the *E. coli* ribosomal protein S1[21-22] contains an OB (Oligonucleotide/oligosaccharide Binding) fold with a five-stranded β-sheet coiled to form a closed β-barrel (Figs. 3E,F). A typical OB domain binds nucleic acid through surfaces composed of positively charged and hydrophobic residues on the solvent exposed β-sheet (Figs. 3E,F). For instance, the RNase E S1 domain binds polymeric single-stranded nucleic acids via a positively charged surface that comprises strands β2 and β3 and the loops between strands β2 and β3 and strands β3 and β5.[23] Both the KH and S1 domains of PNPase contribute to RNA binding, as simultaneous deletion of both domains impairs the apparent affinity of PNPase for RNA substrates.[24-25] Interestingly, the orientation of these domains situates the canonical RNA binding surfaces of the KH domain toward the central pore while the putative RNA binding surfaces of the S1 domain face outward near the exterior of the complex.

Archaeal Csl4 contains three domains: the NTD, S1 domain and CTD. The NTD consists of 2 symmetrical three stranded β-sheets and the CTD contains a 3-stranded β-sheet that coordinates a Zn^{2+} via four cysteine residues that is similar to the iron-binding portion of rubredoxins (Fig. 3E).[26-27] Eukaryotic Csl4 shares structural similarity to archaeal Csl4 and contains an NTD, S1 domain and CTD; however, despite having a similar rubredoxin-like fold, the four cysteine residues in the CTD that coordinate zinc in archaeal Csl4 are not conserved in eukaryotes.[9] Archaeal Rrp4 contains three domains: the NTD, KH Type I domain and a C-terminal S1 domain.[6] Eukaryotic Rrp4 and Rrp40 share structural similarity to archaeal Rrp4 and each contains an NTD, a central KH Type I domain and a C-terminal S1 domain; however both subunits lack the canonical GXXG motif in their KH domains that is believed to be important for RNA interactions.

The arrangement of Csl4 and Rrp4 subunits on the six-subunit ring in the archaeal exosome positions the positively charged putative RNA binding S1 domain surfaces facing toward the central pore while the NTD and KH domains are position nearer to the periphery of the complex (Fig. 2C). It remains unclear how these domains interact with RNA. For instance, Rrp4 from *S. solfataricus* promotes interactions with a poly(A) RNA substrate in the context of the exosome, as evidenced by its ability to increase the affinity for RNA by ~30 fold compared to the archaeal Rrp41-Rrp42 six-component ring alone.[28] However, X-ray structures of archaeal exosomes bound to RNA have so far only elucidated interactions between RNA substrates and residues within the central pore of the six-component Rrp41-Rrp42 ring, despite the presence of the three-component cap.[7] Eukaryotic Rrp4, Rrp40 and Csl4 subunits are similarly positioned on the human exosome core,[9,28] directing putative RNA binding surfaces of the respective S1 domains toward the central channel and the putative RNA binding surfaces of the KH domains toward the periphery of the complex. Additional experiments will be required to characterize the relevance of the putative RNA binding surfaces in the three-subunit exosome cap.

MECHANISM OF PHOSPHOROLYTIC ACTIVITY IN BACTERIAL PNPase AND ARCHAEAL EXOSOMES

Bacterial PNPase and archaeal exosomes contain three identical active sites within the central pore that catalyze the phosphorolytic 3′ to 5′ exoribonuclease activity (Figs. 2B,C).[5,8,29] By analysis of the RNA-free and RNA-bound X-ray structures of PNPase, it was determined that the alpha domain, which partially obstructs the bottom entrance to the central pore, can transition from partially disordered to ordered upon RNA coordination.[2,8] In addition, two narrow constrictions in the pore are believed to regulate access to the phosphorolytic active sites. The first lies near the entrance to the central pore and features three phenylalanine side chains (Fig. 3B), one from each PNPase protomer, that each base stack with one nucleotide (presumably from three different RNA oligomers). The second constriction is located deep within the PNPase central channel near the active site (Fig. 3B).

Archaeal exosomes also recruit RNA to the active site via interactions with at least two RNA binding surfaces that reside in the Rrp41-Rrp42 heterodimer within the central pore (Fig. 3C). Interactions between the two RNA binding surfaces and RNA have been observed for X-ray crystal structures of archaeal exosomes using poly(A) or poly(U) RNA polymers of varying length.[7,17-18] The first interaction surface is located within a loop at the top of the central pore near the three-component cap interface and it features a histidine residue from Rrp41 that stacks with a nitrogenous base near the 5′ end of the RNA. The second RNA interaction surface is proximal to the phosphorolytic active site and includes extensive contacts to the RNA substrate; this surface forms a 10 Å constriction of the central pore and thus it is believed to allow only one RNA molecule to pass through the pore at one time. Protein contacts to the RNA include ribose specific interactions at the 3′OH terminal nucleotide and contacts to the fourth to last nucleotide position through phosphate backbone interactions and nitrogenous base stacking interactions (Fig. 4C). No electron density has yet been observed for RNA nucleotides between these two RNA binding surfaces, thus it has been speculated that the intervening RNA nucleotides are not coordinated in any particular configuration.[7,15,17]

Structural insight into the catalytic mechanism during phosphorolysis can be gleaned by comparing active sites from a variety of X-ray crystal structures of bacterial PNPase and archaeal exosomes in complex with different ligands (Fig. 4). X-ray structures of PNPase with either manganese cations or tungstate revealed the identity of active site residues that coordinate magnesium or phosphate (Fig. 4A), respectively.[8] Residues that coordinate the phosphate include Ser437, Ser438, Ser439 and key residues that coordinate the magnesium include Asp486 and Asp492. A phosphate-binding site composed of similar amino acid side chains was also deduced in an analogous position for the *S. solfataricus* archaeal exosome through identification of a chloride ion[7] and for the *A. fulgidus* exosome through identification of a tungstate ion.[6] Structures of the *S. solfataricus* archaeal exosome in complex with poly (A) RNA and the ADP product (Fig. 4C,D) further identified residues that coordinate the phosphate of the NDP product and phosphodiester backbone of the RNA substrate (Arg99 and Arg139).[17]

A composite active site can be extrapolated from these structures onto the PNPase structure to provide a structural rationale for the phosphorolytic reaction mechanism (Fig. 4E). Two serine residues position a phosphate ion proximal to the phosphodiester linkage between the terminal and penultimate nucleotides. The magnesium ion, His403 and Lys494 position the terminal bridging phosphate in the substrate in an appropriate

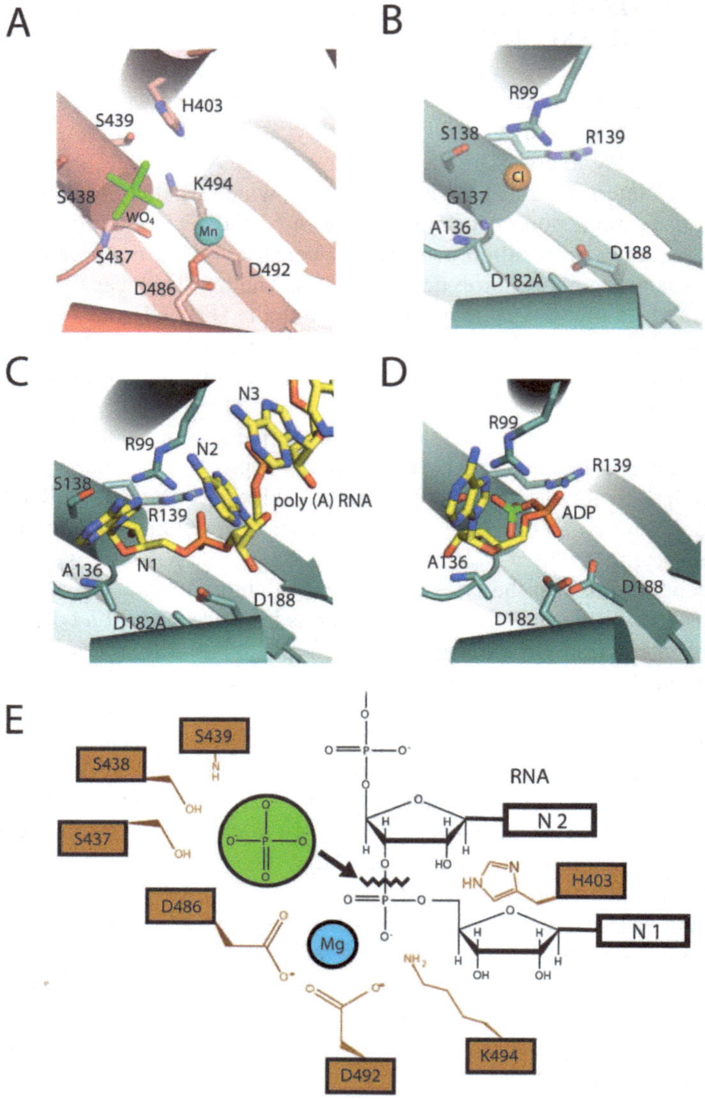

Figure 4. Phosphorolytic exoribonuclease catalytic mechanism for PNPase and the archaeal exosome. A) PNPase active site. A composite structure is depicted that shows the phosphate-mimic tungstate (green) and magnesium ion-mimic manganese (blue sphere) (PDB ID = 1E3P and 3GME). Residues that bind "phosphate" include: S439, S438 and S437. Residues that coordinate "magnesium" are D486 and D492. B) *S. solfataricus* exosome active site. The phosphate-mimic chloride is shown as a yellow sphere and residues that bind the proposed phosphate-mimic chloride are: A136, G137 and S138 (PDB ID = 2BR2). C) Left panel: D182A mutant *S. solfataricus* in complex with a five-nucleotide poly (A) RNA substrate (PDB ID = 2C38). Nucleotides are colored yellow and numbered in such a manner that the first nucleotide (N1) is at the 3′OH end. (D) *S. solfataricus* in complex with the product ADP (PDB ID = 2C39). ADP is colored in yellow with the α-phosphate colored orange and β-phosphate colored green. E) Proposed phosphorolytic exoribonuclease mechanism. Ser437 and Ser439 provide a binding pocket for phosphate (green). Asp486 and Asp492 coordinate a magnesium ion. The magnesium, with the aid of K494 and H403, positions the bridging phosphate between N1 and penultimate N2 nucleotides to facilitate in-line attack by the phosphate.

configuration to facilitate in-line attack by the phosphate nucleophile, which ultimately results in the formation of the NDP product. Although *S. solfataricus* Asp182 and Asp188 are predicted to be required for metal coordination based on sequence similarity, no metal ion has yet been observed in the active sites of *S. solfataricus* exosomes.

Rrp44, A EUKARYOTIC EXOSOME SUBUNIT WITH HYDROLYTIC ENDORIBONUCLEASE AND PROCESSIVE EXORIBONUCLEASE ACTIVITIES

As discussed earlier, none of the human or budding yeast subunits that comprise the 9-component exosome core retain phosphorolytic exoribonuclease activity because most of the key active site residues required for RNA binding or for metal and phosphate coordination have not been conserved across evolution.[9-10,30] Studies with the budding yeast exosome have demonstrated that the tenth exosome subunit, Rrp44 (also known as Dis3), is solely responsible for the processive hydrolytic activity that is associated with the exosome in the cytoplasm.[9-10,31] It is important to note that while human encodes three apparent homologs of budding yeast Rrp44, human Rrp44 has not yet been shown to associate with the human exosome core.[9,32] Rrp44 exoribonuclease activity results in hydrolysis of RNA one nucleotide at a time in a 3′ to 5′ direction, releasing 5′ nucleotide monophosphates in a sequence independent manner.[30]

Rrp44 contains five domains: an endoribonucleolytic active site containing PIN (PIlus-forming N-terminus) domain, two cold shock domains (CSD1 and CSD2), a central hydrolytic exoribonucleolytic active site containing domain (RNB) and an S1 domain (Fig. 1C).[31] The overall architecture of Rrp44 has been determined based on two structures of Rrp44: one determined in complex with RNA in the absence of the PIN domain and one for full-length Rrp44 in complex with Rrp41 and Rrp45. These structures reveal the modular architecture of Rrp44 in which the PIN domain is located above the two CSDs and S1 domain with the RNB domain located below the two CSDs and S1 domain (Fig. 5A).[19,30]

Rrp44 is structurally and mechanistically related to bacterial RNase II and RNase R, however comparison of Rrp44 to structures of RNase II in apo- and RNA-bound states reveals that RNase II has a slightly different arrangement of the cold shock domains and S1 domain (Fig. 5B).[33] An Rrp44-RNA complex showed that CSD1 engages in interactions with the RNA substrate to guide it into the RNB domain exoribonucleolytic active site (Fig. 5A, right panel) and interactions between the single stranded RNA substrate and CSD1 facilitate recruitment to the exoribonucleolytic catalytic site by a specific orientation of the three OB-containing domains (CSD1, CSD2 and S1). In comparison, RNase II positions the three OB-containing domains in a different conformation that allows for single stranded RNA interactions with CSD2 and the S1 domains (Fig. 5B, right panel). As will be discussed, these two alternative modes of RNA interaction present fundamentally different paths that serve to guide the RNA to the exoribonucleolytic active site.

Rrp44 PIN DOMAIN

The PIN domain family, named after its apparent homology with the N-terminal domain of the pili biogenesis protein detected in some bacteria, includes over 300 members

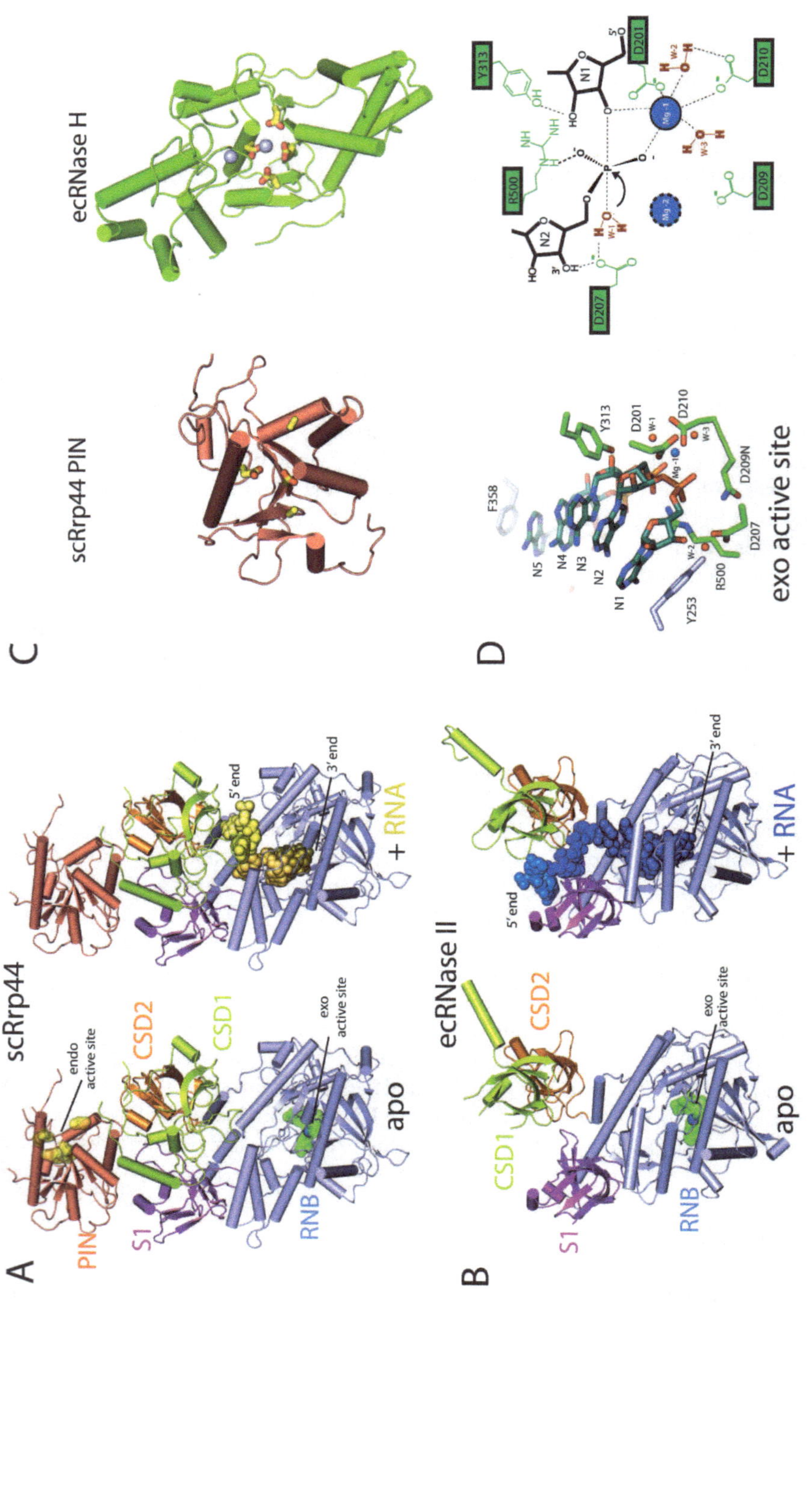

Figure 5. Please see figure legend on following page.

Figure 5, viewed on previous page. Eukaryotic Rrp44 structure and catalytic mechanism. A) Structures of *S. cerevisiae* Rrp44 with and without poly(A) RNA. The Rrp44 domains are PIN (pink), CSD1 (lime), CSD2 (orange), RNB (blue) and S1 (purple) (PDB ID = 2WP8). Residues D91, E120, D171 and D198 are colored yellow and indicate the position of the endoribonucleolytic active site. Residues D543, D540, D551N and D552 are colored green and indicate the position of the exoribonucleolytic active site. RNA (yellow spheres) was modeled into the full length Rrp44 (right panel) by alignment to the poly(A) RNA bound Rrp44ΔPIN structure (PDB ID = 2VNU). B) Structures of *E. coli* RNase II with and without poly(A) RNA. RNase II domains are CSD1 (lime), CSD2 (orange), RNB (blue) and S1 (purple). Residues D201, D207, D209 and D210 are colored green and indicate the position of the exoribonucleolytic active site. The magnesium ion is shown as a blue sphere (PDB ID = 2IXO). The poly(A) RNA (blue) bound structure of RNase II is shown in a similar orientation in the right panel (PDB ID = 2IX1). C) Structures shown for the *S. cerevisiae* PIN from Rrp44 (left, PDB ID = 2WP8) and *E. coli* bacteriophage T4 RNase H (right, PDB ID = 1TFR). Active site residues are highlighted in yellow and magnesium ions shown as blue spheres. D) The hydrolytic exoribonucleolytic active site of RNase II. Left panel: Structural representation of relevant residues that coordinate magnesium ions, Mg-1 and Mg-2 (not detected) and residues that coordinate RNA are shown in green. Y253 and F358 are shown in blue and make base stacking interactions with the RNA substrate. Water molecules (W-1, W-2 and W-3), important for the reaction mechanism, are shown as red spheres. Right panel: Schematic representation of the active site. Representation depicts the binding of W-1 by magnesium ions and charged residues to facilitate nucleophilic attack of the bridging phosphate between nucleotides N1 and N2.

found in bacteria, archaea and eukaryotes.[34] The PIN domain consists of a central twisted five-stranded β sheet flanked by α helices and an active site that is capable of cleaving nucleic acid (Fig. 5C, left panel). Biophysical and structural studies of PAE2754 from *Pyrobaculum aerophilum* and the OT3 protein from *Pyrococcus horkoshii* revealed that PIN domains can form dimers and dimers of dimers, respectively.[35-36] While the precise nature of the binding surfaces required for recruiting nucleic acids to the PIN domain active site remain unknown, it was postulated for PAE2754 that nucleotides thread through a central channel that is formed via tetramerization. Although many structures of PIN domains reveal dimers within the crystal lattice, it remains uncertain if dimerization is a functionally relevant feature for all catalytically active PIN domains.

The nucleolytic active site, detected within some but not all PIN domains,[37] consists of four conserved acidic residues that coordinate two divalent cations. For the PIN domain in budding yeast Rrp44, all four acidic residues are present. In higher eukaryotes such as humans, the three Rrp44 paralogs vary with respect to the conservation of the acidic residues or the presence of an intact PIN domain (Fig. 1C). The functional consequence of this variation within higher eukaryotes has yet to be investigated. Structures of the PIN domains also reveal a striking structural homology to T4 RNase H, despite a dearth in sequence identity (Fig. 5C) and therefore the endoribonucleolytic activity of PIN domains is predicted to utilize a similar two metal-dependent catalytic mechanism for hydrolysis of nucleotides.[38-39]

Rrp44 RNB DOMAIN

3′ to 5′ exoribonuclease activity is catalyzed in an active site within the RNB domain which is located at the end of a narrow channel and composed by acidic residues that coordinate two divalent metal ions, motifs conserved in bacterial RNase II and RNase R. Analysis of the RNase II apo- and RNA bound structures provide insights to the catalytic mechanism of exoribonucleolytic decay.[33] The active site of RNase II is composed by acidic residues Asp201, Asp207, Asp209 and Asp210 that coordinate two magnesium ions (Fig. 5D).[33] The first (Mg-1) is coordinated by residues Asp201 and Asp210 and two waters, W-2 and W-3. The second (Mg-2) was not detected in X-ray structures of RNase II, but is speculated to be coordinated by Asp207 and Asp209. Interestingly, electron density for the Arg500 side chain was only detected when in the presence of RNA where it was observed coordinating the phosphate bridging the last two 3′ nucleotides (N1 and N2), conceivably to stabilize the transition state during cleavage of the phosphodiester bond. The exoribonucleolytic reaction is believed to proceed through a two-metal-ion bimolecular nucleophilic substitution mechanism. W-1, coordinated by Asp207 and Mg-2, is the nucleophile for in-line attack of the phosphate between the last and penultimate nucleotide, ultimately resulting in release of the NMP product.

A 'clamp' action is thought to promote translocation between successive cleavage events, thus leading to the processive degradation activities observed for this enzyme family. RNase II achieves this function by employing base stacking interactions between N5 and F358 as well as N1 and Y253 to stabilize the RNA substrate within the active site.[33] A similar mechanism for catalysis and translocation can be inferred for budding yeast Rrp44 by comparing the conserved active site residues of the RNA bound Rrp44ΔPIN X-ray crystal structure.[30] It should be noted that RNase II is unable to process structured RNA substrates, unlike the bacterial RNase R and eukaryotic Rrp44.[9,19,30,40-41]

Rrp44 AND THE 10-COMPONENT EXOSOME

Models for the structure of the eukaryotic 10-component exosome have been proposed based on the X-ray structure of the human nine-component exosome core (Liu et al, 2006), the X-ray structure of the budding yeast Rrp41-Rrp45-Rrp44 trimer[19] and a 20 Å resolution negative-stain EM structure of budding yeast Rrp44-bound to the core exosome (in the absence of Csl4).[42] In the case of the Rrp41-Rrp45-Rrp44 trimer, the ten-component exosome was modeled by aligning the budding yeast Rrp41-Rrp45 proteins to the respective human counterparts (Fig. 6).[19] In the case of the EM structure, the human exosome core structure was positioned in the EM density followed by docking the Rrp44 RNB domain into remaining density. In each of these models, the PIN domain of Rrp44 interacts with the 'bottom' of the exosome core principally through interactions with Rrp41 and Rrp45. Although these models are in general agreement, additional interactions have been reported that include surfaces identified in the EM structure between the CSD1 of Rrp44 and Rrp43[42] and protomer interactions between Rrp44, Rrp41, Rrp45, Rrp42 and Rrp4, detected by the presence of sub-complexes of budding yeast exosomes by mass spectrometry experiments.[43] However, it should be noted that Rrp42 and Rrp4 may interact indirectly with Rrp44 through Rrp41 and Rrp45. Based on analysis of the architecture of the 10-component exosome and comparisons to the RNA bound Rrp44-ΔPIN and RNA bound RNase II structures, it was predicted that RNA threads into the exosome core through the three-component cap, progressing through the central pore of the exosome to direct the 3′ OH end of the RNA substrate into the hydrolytic exoribonucleolytic active site of the Rrp44 RNB domain (Fig. 6).

As discussed in previous sections, the path of the RNA substrate into the exoribonucleolytic active site differs significantly in structures of budding yeast Rrp44 and bacterial RNase II despite conservation of the CSD and S1 domains. Extrapolating the path of RNA observed in these structures within the context of a 10-component exosome model, RNA would be required to exit the bottom of the 9-component exosome, become solvent exposed and would then be required to make a ~45° turn around CSD1 of Rrp44 in order to enter into the Rrp44 channel that leads to the RNB active site (Fig. 6). While this route appears circuitous, it is consistent with both RNase protection and RNA exoribonuclease decay assays which indicated that RNA substrates required inclusion of at least 31-34 single stranded nucleotides at the 3′ end to be engaged by the 10-component exosome.[19] However, it is also conceivable that RNA binding could induce conformational changes in the complex to facilitate a more direct path for RNA substrates into the Rrp44 active site.

In contrast to the RNA path predicted for the 10-component exosome based on the RNA bound Rrp44-ΔPIN structure, the path predicted for RNA based on the structure of an RNase II-RNA complex, would place the RNA perpendicular to the central pore of the 9-component exosome (Fig. 6). This model is not consistent with the utilization of the central pore for RNA interactions but is consistent with a role for the PIN domain in RNA interactions, as this RNA path, if extended, points directly toward the PIN domain. Taken together, these structural models suggest that alternative binding modes may exist to engage the RNB domain of Rrp44 either via interactions with the PIN or the central pore of the exosome.

Rrp6, A EUKARYOTIC EXOSOME SUBUNIT WITH DISTRIBUTIVE HYDROLYTIC ACTIVITIES

Rrp6 contains at least three domains: an N-terminal domain (NTD), an exoribonuclease domain (EXO) that contains the DEDD-Y active site amino acid motif detected in many DNA and RNA nucleases and a Helicase and RNase D Carboxy terminal (HRDC) domain (Fig. 1C).[44] The structure of a catalytically active domain of budding yeast Rrp6 shares structural homology to RNase D from *E. coli* (Fig. 7).[45] RNase D contains the EXO domain with a DEDD-Y active site, but differs in that it contains two flanking HRDC domains that together form a funnel shaped ring. The two HRDC domains in RNase D were proposed to recruit RNA substrates, channeling them to the active site for processing. While a second HRDC domain has not been identified in the ~200 C-terminal residues in eukaryotic Rrp6, a similar hypothesis for RNA binding and recruitment to the active site has been suggested for the HRDC domain based on sequence similarity to the RecQ helicase protein family and the fact that this domain is critical for processing RNAs such as 5.8 S rRNA and snR40 snoRNA, as determined using Rrp6 isolated from *S. cerevisiae*.[46]

The DEDD active site is observed in a variety of nucleolytic enzymes that catalyze degradation of DNA and RNA as exemplified by the Klenow fragment of DNA polymerase I.[47] A two-metal assisted catalytic mechanism has been proposed based on X-ray crystal structures and mutational analysis, in which the negatively charged DEDD residues coordinate two metal ions that are required for cleavage of the phosphodiester bond.[48] A similar model has been proposed for RNase D and Rrp6: a magnesium ion acts as a Lewis acid to deprotonate a water molecule and then the phosphodiester backbone is attacked by the resulting nucleophilic water at the penultimate nucleotide of the RNA substrate. The DEDD-Y active site of RNase D, Rrp6 and related enzymes is unique compared to other enzymes containing DEDD active sites, because they employ an additional tyrosine proximal to the DEDD active site to coordinate the nucleophilic water. The distributive 3′ to 5′ exoribonuclease activity observed for Rrp6 is consistent with the structure because unlike Rrp44, whose active site is sequestered at the end of

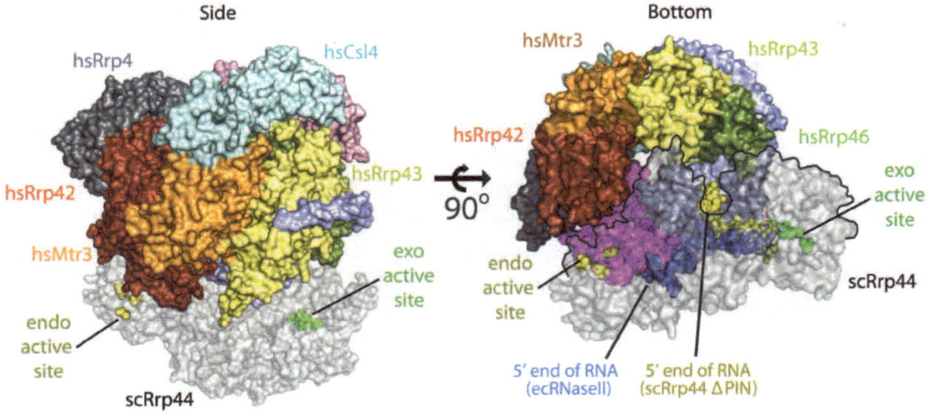

Figure 6. Model for RNA recruitment to the hydrolytic active site of Rrp44 within the eukaryotic exosome. A 10-component exosome model was created by aligning the *S. cerevisiae* Rrp41-Rrp45-Rrp44 trimer (PDB ID = 2WP8) onto the Rrp41-Rrp45 subunits of the human exosome (PDB ID = 2NN6). Coloring for the 9-component exosome is described in Figure 2 and the Rrp44 component is shaded grey. The left panel depicts a side view of the complex with the Rrp44 exoribonucleolytic active site indicated as green spheres and the endoribonucleolytic active site as yellow spheres. The right panel depicts a bottom view of the complex. The RNA complexes determined for RNase II (blue spheres, PDB ID = 2IX1) and Rrp44ΔPIN (yellow spheres, PDB ID = 2VNU) were superimposed into the full-length Rrp44 structure to illustrate the paths of RNA in the complex. The Rrp44 molecule is outlined by a black line in the right panel where it overlaps with the exosome core.

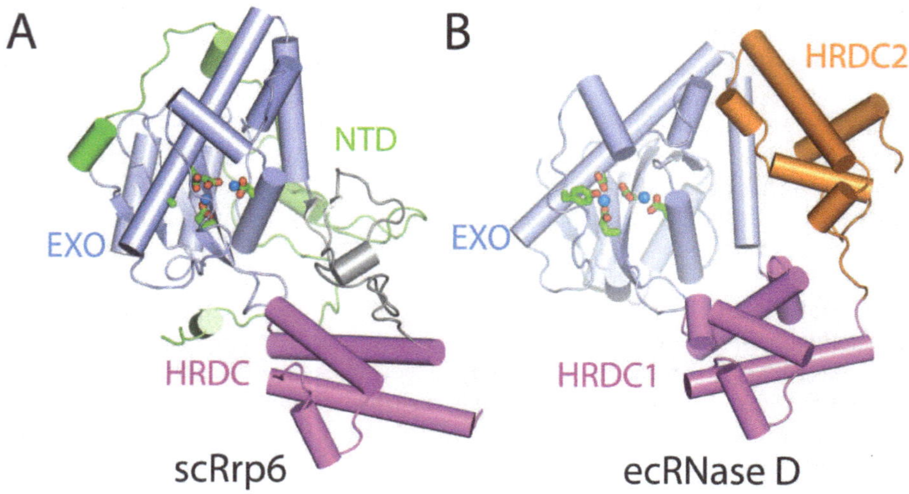

Figure 7. Structure of eukaryotic Rrp6 and the bacterial homolog RNase D. A) *S. cerevisiae* Rrp6 structure (PDB ID = 2HBL). Domains of Rrp6 are NTD (green), EXO (blue) and HRDC (pink). The active site residues (D238, E240, D296, D365 and Y361A) are shown coordinating two manganese ions (blue spheres). B) *E. coli* RNase D structure (PDB ID = 1YT3). Domains of RNase D are EXO (blue), HRDC1 (pink) and HRDC2 (orange). The active site residues (D28, E30, D85, D155 and Y151) colored green are coordinating two zinc ions (blue spheres).

a deep channel, the Rrp6 active site is exposed on the surface of the enzyme. Further mechanistic insight to Rrp6 interactions with RNA substrates will require additional structures and biochemical analysis of Rrp6 complexes with RNA substrates as none are yet resolved.

Rrp6 INTERACTIONS WITH THE EXOSOME CORE

No detailed atomic resolution structures exist for Rrp6 in association with the exosome core, although a 35 Å resolution negative-stain EM structure of the *L. tarentolae* exosome core has been determined in complex with Rrp6 and Rrp47, an accessory protein that was reported to increase the exoribonuclease activity of the exosome.[49] From this work, the authors proposed a model whereby Rrp6 and Rrp47 interact with the 9-component exosome core near the 'top' and adjacent to the three-component cap. While this organization may apply to the *L. tarentolae* exosome, it remains unclear if this organization will apply to other eukaryotic exosomes because yeast two-hybrid data demonstrated that human Rrp6 interacts with Rrp41, Rrp43, Rrp46 and Mtr3 suggesting that Rrp6 may also interact with the six-subunit ring of the exosome.[50]

Unlike the exoribonucleolytic activities of Rrp44 which are clearly modulated or regulated via association with the exosome core,[9,19] similar activities were observed in vitro for budding yeast Rrp6 prior to and after its association with the exosome core.[9] With that said, it is clear that Rrp6 association with the exosome core is important for targeting Rrp6 to its physiological substrates, as evidenced by the fact that a fragment of Rrp6 that loses its ability to interact with the core (but retains catalytic function) is not sufficient to complement many of the functions of Rrp6 in vivo.[51] However, it is also interesting that the activities of Rrp6 can be stimulated without the core exosome by association with members of the TRAMP complex, in a manner independent of the Trf4 poly(A) polymerase and Mtr4 RNA helicase activities.[52] Further investigations will be required to determine the structural basis for these seemingly disparate activities.

CONCLUSION

Structures and models derived for exosomes from bacteria, archaea and eukaryotes demonstrate a striking architectural similarity with respect to 1) the six RNase PH-like domains that oligomerize to form a pseudo-hexameric ring and 2) the orientation of the S1 and KH RNA binding domains that form a trimeric cap on top of the exosome. This structural framework results in formation of a central channel. In bacterial PNPase and archaeal exosomes, this channel harbors RNA binding surfaces and phosphorolytic active sites and because the central channel is narrow, only single stranded RNAs can thread through the central pore via interactions with two conserved RNA binding surfaces. Furthermore, the two RNA binding surfaces confer processivity during RNA decay, presumably by preventing RNA substrates from diffusing away from the complex between successive rounds of cleavage.

Eukaryotic exosomes have been reported to use the same strategy to engage RNA substrates by utilizing the inactive 9-component exosome core to bind and transport

RNA substrates through the pore to ultimately engage the hydrolytic exoribonuclease activities of Rrp44, although this has not been demonstrated in any structural detail. It also remains unclear how Rrp6 engages the exosome core and whether it too is influenced by the RNA binding properties of the exosome core channel. On a final note, it is known that the RNA exosome interacts with several other factors including the TRAMP and SKI complexes among others,[53-55] suggesting that additional surfaces of the exosome core may be required for recruitment of these effectors to alter or regulate exosome activity. Although much has been accomplished since the discovery of the eukaryotic exosome more than ten years ago,[56] it is clear that much work remains to fully understand how the molecular architecture of the eukaryotic exosome impacts on its biochemical and cellular functions.

ACKNOWLEDGEMENTS

K.J. and C.D.L. are supported in part by a grant from the National Institutes of Health (GM079196).

REFERENCES

1. Carpousis AJ. The Escherichia coli RNA degradosome: structure, function and relationship in other ribonucleolytic multienzyme complexes. Biochem Soc Trans 2002; 30(2):150-155.
2. Symmons MF, Jones GH, Luisi BF. A duplicated fold is the structural basis for polynucleotide phosphorylase catalytic activity, processivity and regulation. Structure 2000; 8(11):1215-1226.
3. Evguenieva-Hackenberg E, Walter P, Hochleitner E et al. An exosome-like complex in Sulfolobus solfataricus. EMBO Rep 2003; 4(9):889-893.
4. Koonin EV, Wolf YI, Aravind L. Prediction of the archaeal exosome and its connections with the proteasome and the translation and transcription machineries by a comparative-genomic approach. Genome Res 2001; 11(2):240-252.
5. Lorentzen E, Walter P, Fribourg S et al. The archaeal exosome core is a hexameric ring structure with three catalytic subunits. Nat Struct Mol Biol 2005; 12(7):575-581.
6. Buttner K, Wenig K, Hopfner KP. Structural framework for the mechanism of archaeal exosomes in RNA processing. Mol Cell 2005; 20(3):461-471.
7. Lorentzen E, Dziembowski A, Lindner D et al. RNA channelling by the archaeal exosome. EMBO Rep 2007; 8(5):470-476.
8. Nurmohamed S, Vaidialingam B, Callaghan AJ et al. Crystal structure of Escherichia coli polynucleotide phosphorylase core bound to RNase E, RNA and manganese: implications for catalytic mechanism and RNA degradosome assembly. J Mol Biol 2009; 389(1):17-33.
9. Liu Q, Greimann JC, Lima CD. Reconstitution, activities and structure of the eukaryotic RNA exosome. Cell 2006; 127(6):1223-1237.
10. Dziembowski A, Lorentzen E, Conti E et al. A single subunit, Dis3, is essentially responsible for yeast exosome core activity. Nat Struct Mol Biol 2007; 14(1):15-22.
11. Liu QS, Greimann JC, Lima CD. Reconstitution, activities and structure of the eukaryotic RNA exosome (vol 127, pg 1223, 2006). Cell 2007; 131(1):188-189.
12. Shi Z, Yang WZ, Lin-Chao S et al. Crystal structure of Escherichia coli PNPase: central channel residues are involved in processive RNA degradation. RNA 2008; 14(11):2361-2371.
13. Harlow LS, Kadziola A, Jensen KF et al. Crystal structure of the phosphorolytic exoribonuclease RNase PH from Bacillus subtilis and implications for its quaternary structure and tRNA binding. Protein Sci 2004; 13(3):668-677.
14. Ishii R, Nureki O, Yokoyama S. Crystal structure of the tRNA processing enzyme RNase PH from Aquifex aeolicus. J Biol Chem 2003; 278(34):32397-32404.
15. Navarro MV, Oliveira CC, Zanchin NI et al. Insights into the mechanism of progressive RNA degradation by the archaeal exosome. J Biol Chem 2008; 283(20):14120-14131.

16. Stehmeier P, Muller S. Phospho-regulated SUMO interaction modules connect the SUMO system to CK2 signaling. Mol Cell 2009; 33(3):400-409.
17. Lorentzen E, Conti E. Structural basis of 3′ end RNA recognition and exoribonucleolytic cleavage by an exosome RNase PH core. Mol Cell 2005; 20(3):473-481.
18. Ramos CR, Oliveira CL, Torriani IL et al. The Pyrococcus exosome complex: structural and functional characterization. J Biol Chem 2006; 281(10):6751-6759.
19. Bonneau F, Basquin J, Ebert J et al. The yeast exosome functions as a macromolecular cage to channel RNA substrates for degradation. Cell 2009; 139(3):547-559.
20. Valverde R, Edwards L, Regan L. Structure and function of KH domains. FEBS J 2008; 275(11):2712-2726.
21. Bycroft M, Hubbard TJ, Proctor M et al. The solution structure of the S1 RNA binding domain: a member of an ancient nucleic acid-binding fold. Cell 1997; 88(2):235-242.
22. Subramanian AR. Structure and functions of ribosomal protein S1. Prog Nucleic Acid Res Mol Biol 1983; 28:101-142.
23. Schubert M, Edge RE, Lario P et al. Structural characterization of the RNase E S1 domain and identification of its oligonucleotide-binding and dimerization interfaces. J Mol Biol 2004; 341(1):37-54.
24. Briani F, Curti S, Rossi F et al. Polynucleotide phosphorylase hinders mRNA degradation upon ribosomal protein S1 overexpression in Escherichia coli. RNA 2008; 14(11):2417-2429.
25. Stickney LM, Hankins JS, Miao X et al. Function of the conserved S1 and KH domains in polynucleotide phosphorylase. J Bacteriol 2005; 187(21):7214-7221.
26. Blake PR, Park JB, Zhou ZH et al. Solution-state structure by NMR of zinc-substituted rubredoxin from the marine hyperthermophilic archaebacterium Pyrococcus furiosus. Protein Sci 1992; 1(11):1508-1521.
27. Day MW, Hsu BT, Joshua-Tor L et al. X-ray crystal structures of the oxidized and reduced forms of the rubredoxin from the marine hyperthermophilic archaebacterium Pyrococcus furiosus. Protein Sci 1992; 1(11):1494-1507.
28. Oddone A, Lorentzen E, Basquin J et al. Structural and biochemical characterization of the yeast exosome component Rrp40. EMBO Rep 2007; 8(1):63-69.
29. Jarrige A, Brechemier-Baey D, Mathy N et al. Mutational analysis of polynucleotide phosphorylase from Escherichia coli. J Mol Biol 2002; 321(3):397-409.
30. Lorentzen E, Basquin J, Tomecki R et al. Structure of the active subunit of the yeast exosome core, Rrp44: diverse modes of substrate recruitment in the RNase II nuclease family. Mol Cell 2008; 29(6):717-728.
31. Schneider C, Anderson JT, Tollervey D. The exosome subunit Rrp44 plays a direct role in RNA substrate recognition. Mol Cell 2007; 27(2):324-331.
32. Chen CY, Gherzi R, Ong SE et al. AU binding proteins recruit the exosome to degrade ARE-containing mRNAs. Cell 2001; 107(4):451-464.
33. Frazao C, McVey CE, Amblar M et al. Unravelling the dynamics of RNA degradation by ribonuclease II and its RNA-bound complex. Nature 2006; 443(7107):110-114.
34. Wall D, Kolenbrander PE, Kaiser D. The Myxococcus xanthus pilQ (sglA) gene encodes a secretin homolog required for Type IV pilus biogenesis, social motility and development. J Bacteriol 1999; 181(1):24-33.
35. Arcus VL, Backbro K, Roos A et al. Distant structural homology leads to the functional characterization of an archaeal PIN domain as an exonuclease. J Biol Chem 2004; 279(16):16471-16478.
36. Jeyakanthan J, Inagaki E, Kuroishi C et al. Structure of PIN-domain protein PH0500 from Pyrococcus horikoshii. Acta Crystallogr Sect F Struct Biol Cryst Commun 2005; 61(Pt 5):463-468.
37. Glavan F, Behm-Ansmant I, Izaurralde E et al. Structures of the PIN domains of SMG6 and SMG5 reveal a nuclease within the mRNA surveillance complex. EMBO J 2006; 25(21):5117-5125.
38. De Vivo M, Dal Peraro M, Klein ML. Phosphodiester cleavage in ribonuclease H occurs via an associative two-metal-aided catalytic mechanism. J Am Chem Soc 2008; 130(33):10955-10962.
39. Steitz TA, Steitz JA. A general two-metal-ion mechanism for catalytic RNA. Proc Natl Acad Sci USA 1993; 90(14):6498-6502.
40. Cheng ZF, Deutscher MP. Purification and characterization of the Escherichia coli exoribonuclease RNase R. Comparison with RNase II. J Biol Chem 2002; 277(24):21624-21629.
41. Cheng ZF, Deutscher MP. An important role for RNase R in mRNA decay. Mol Cell 2005; 17(2):313-318.
42. Wang HW, Wang J, Ding F et al. Architecture of the yeast Rrp44 exosome complex suggests routes of RNA recruitment for 3′ end processing. Proc Natl Acad Sci USA 2007; 104(43):16844-16849.
43. Hernandez H, Dziembowski A, Taverner T et al. Subunit architecture of multimeric complexes isolated directly from cells. EMBO Rep 2006; 7(6):605-610.
44. Midtgaard SF, Assenholt J, Jonstrup AT et al. Structure of the nuclear exosome component Rrp6p reveals an interplay between the active site and the HRDC domain. Proc Natl Acad Sci USA 2006; 103(32):11898-11903.

45. Zuo Y, Wang Y, Malhotra A. Crystal structure of Escherichia coli RNase D, an exoribonuclease involved in structured RNA processing. Structure 2005; 13(7):973-984.
46. Phillips S, Butler JS. Contribution of domain structure to the RNA 3′ end processing and degradation functions of the nuclear exosome subunit Rrp6p. RNA 2003; 9(9):1098-1107.
47. Ollis DL, Brick P, Hamlin R et al. Structure of large fragment of Escherichia coli DNA polymerase I complexed with dTMP. Nature 1985; 313(6005):762-766.
48. Beese LS, Steitz TA. Structural basis for the 3′-5′ exonuclease activity of Escherichia coli DNA polymerase I: a two metal ion mechanism. EMBO J 1991; 10(1):25-33.
49. Cristodero M, Bottcher B, Diepholz M et al. The Leishmania tarentolae exosome: purification and structural analysis by electron microscopy. Mol Biochem Parasitol 2008; 159(1):24-29.
50. Lehner B, Sanderson CM. A protein interaction framework for human mRNA degradation. Genome Res 2004; 14(7):1315-1323.
51. Callahan KP, Butler JS. Evidence for core exosome independent function of the nuclear exoribonuclease Rrp6p. Nucleic Acids Res 2008; 36(21):6645-6655.
52. Callahan KP, Butler JS. TRAMP complex enhances RNA degradation by the nuclear exosome component Rrp6. J Biol Chem 2010; 285(6):3540-3547.
53. Bayne EH, White SA, Allshire RC. DegrAAAded into silence. Cell 2007; 129(4):651-653.
54. Houseley J, Tollervey D. The nuclear RNA surveillance machinery: the link between ncRNAs and genome structure in budding yeast? Biochim Biophys Acta 2008; 1779(4):239-246.
55. Lebreton A, Seraphin B. Exosome-mediated quality control: substrate recruitment and molecular activity. Biochim Biophys Acta 2008; 1779(9):558-565.
56. Mitchell P, Petfalski E, Shevchenko A et al. The exosome: a conserved eukaryotic RNA processing complex containing multiple 3′—>5′ exoribonucleases. Cell 1997; 91(4):457-466.

CHAPTER 3

THE ARCHAEAL EXOSOME

Elena Evguenieva-Hackenberg
Institut für Mikrobiologie und Molekularbiologie, University of Giessen, Giessen, Germany.
Email: Elena.Evguenieva-Hackenberg@mikro.bio.uni-giessen.de

Abstract: The archaeal exosome is a protein complex with structural similarities to the eukaryotic exosome and bacterial PNPase. Its catalytic core is formed by alternating Rrp41 and Rrp42 polypeptides, arranged in a hexameric ring. A flexible RNA binding cap composed of the evolutionarily conserved proteins Rrp4 and/or Csl4 is bound at the top of the ring and seems to be involved in recruitment of specific substrates and their unwinding. Additionally, the protein complex contains an archaea-specific subunit annotated as DnaG, the function of which is still unknown. The archaeal exosome degrades RNA phosphorolytically in 3′ to 5′ direction. In a reverse reaction, it synthesizes heteropolymeric RNA tails using nucleoside diphosphates. The functional similarity between the archaeal exosome and PNPase shows that important processes of RNA degradation and posttranscriptional modification in Archaea are similar to the processes in Bacteria and organelles.

INTRODUCTION

The Archaea are a unique group of prokaryotic micro-organisms also named “the third domain of life”, since they have molecular characteristics that distinguish them from the Bacteria as well as from the Eukarya.[1-3] The best known Archaea are the methanogens and the extremophiles (hyperthermophies, halophiles, acidophilies), but it should be noticed that the vast majority of archaeal species live as ubiquitous mesophiles in water and soil.[3] Archaea are phylogenetically more closely related to Eukarya than to Bacteria,[2] and many of the archaeal proteins and protein complexes are simplified versions of their eukaryotic counterparts.[4,5] Since recombinant protein and protein complexes of hyperthermophilic Archaea can be relatively easily overproduced and crystallized, they are important research objects.

RNA Exosome, edited by Torben Heick Jensen.

Proteins involved in RNA metabolism belong to the evolutionarily most conserved ones and orthologs of several ribonucleases are found in all three domains of life.[6] Koonin at al. proposed the existence of an archaeal exosome in 2001, based on the finding that in most of the sequenced archaeal genomes, three orthologs of eukaryotic exosomal subunits (Rrp4, Rrp41 and Rrp42) are encoded in an array in a highly conserved superoperon.[7] A fourth ortholog of a eukaryotic exosome subunit, Csl4, was found to be encoded in another operon in these genomes. The first experimental evidence for the existence of an exosome-like complex in Archaea was presented in 2003,[8] and in the following years the archaeal exosome was characterized as a phosphorolytic 3′ to 5′ exoribonuclease, which can in an opposite reaction synthesize RNA[9-14] similarly to bacterial polynucleotide phosphorylase (PNPase).[15,16] Here present knowledge of the archaeal exosome is summarized.

CORE SUBUNITS OF THE ARCHAEAL EXOSOME

So far, the in vivo composition of the archaeal exosome was examined for the thermophilic species *Sulfolobus solfataricus*[8,12] and *Methanothermobacter thermoautotrophicus*.[17] The *S. solfataricus* exosome was purified by immunoprecipitation using polyclonal antibodies against the archaeal Rrp41 subunit and coprecipitating proteins were identified by mass spectrometry. The following proteins were found to copurify with Rrp41: Rrp42, Rrp4, Csl4, the archaeal DnaG-like protein, Cpn and a Cdc48 homolog.[8] The exosome of *M. thermoautotrophicus* was separated by Blue Native/SDS-PAGE and mass spectrometry analysis revealed the presence of Rrp41, Rp42, Rrp4, DnaG and the archaeal splicing endonuclease in the complex.[17] In some experiments *S. solfataricus* Csl4 is copurified in very low amounts with the exosome,[12] and this may explain the failure to detect Csl4 in the exosome of *M. thermoautotrophicus*.

It is accepted that Rrp41, Rrp42, Rrp4 and Csl4 belong to the core of the archaeal exosome, since they can be assembled into a complex structurally similar to the eukaryotic nine-subunit exosome.[10,18,19] The archaeal nine-subunit exosome is also structurally and functionally similar to bacterial PNPase,[20] which degrades RNA phosphorolytically but is also responsible for RNA tailing in vivo.[15,16] The nine-subunit form of the archaeal exosome contains three Rrp41, three Rrp42 and three Rrp4 and/or Csl4 polypeptides.[10] Alternating Rrp41 and Rrp42 form the catalytically active hexameric ring,[9,10] on the top of which three Rrp4 and/or Csl4 are located, forming an RNA-binding cap with a central pore.[10,18]

The archaeal DnaG protein was consistently copurified with the exosome and cosediments with Rrp41 and Rrp4 after fractionation of *S. solfataricus* cell-free extracts in glycerol density gradients by ultracentrifugation.[21] Its binding to the exosome is very strong and comparable to the interactions between Rrp41, Rrp42, Rrp4 and Csl4—in coimmunoprecipitation experiments, all five polypeptides still build a complex after washing with 1.8 M NaCl and elute together at low pH.[21] The possibility that *S. solfataricus* DnaG accidentally sticks to the exosome was excluded: it was shown that coimmunoprecipitations with Rrp41-directed or with DnaG-directed antibodies result in purifications of very similar protein complexes and that depletion of Rrp41 from the cell extract is paralleled by DnaG-depletion.[12] Although DnaG strongly interacts with the archaeal exosome and can be considered as its tenth core subunit, the physiological role of the protein remains unknown. As previously discussed,[8,12] its domain composition suggests an RNA-helicase or endoribonuclease function.

The significance of copurification of Cpn and Cdc48 with the *S. solfataricus* exosome is not clear, since both proteins have chaperone properties.[12] Based on the cosedimentation of large amounts of the *S. solfataricus* exosome with 30S and 50S ribosomal subunits in glycerol density gradients, it was proposed that this protein complex is involved in rRNA maturation like its eukaryotic counterpart.[8] The interaction of the *M. thermautotrophicus* exosome with a homomultimer of the splicing endonuclease (bulge-helix-bulge endonuclease) in a 900 kDa complex[17] strongly suggests that the endonuclease and the exosome participate in RNA processing in a coordinate manner.

STRUCTURE AND MECHANISM

The structure and the mechanism of the archaeal exosome were investigated in vitro using protein complexes reconstituted from purified recombinant subunits. Based on early reports about recombinant subunits of the eukaryotic exosome,[22,23] it was expected that at least one of the RNase PH domain (RPD)-containing subunits of the archaeal exosome, Rrp41 or Rrp42, should exhibit phosphorolytic activity and that Rrp4 or Csl4 may show hydrolytic activity. However, none of the recombinant archaeal proteins Rrp41, Rrp42, Rrp4 or Csl4 showed any RNase activity in vitro, when used separately. It turned out that reconstitution of a protein complex containing Rrp41 and Rrp42 is needed for the phosphorolytic activity of the archaeal exosome.[9,12] The activity is modulated in the presence of the RNA-binding subunits Rrp4 and/or Csl4, which do not act as RNases.[10,12] Not only is RNA degradation by the Rrp41/Rrp42 complex more efficient in the presence of Rrp4 and/or Csl4,[10,12,13] RNA binding by Rrp4 is also strongly increased when this protein is part of the exosome,[24] demonstrating the interdependence of the exosomal subunits with respect to the function of the exosome.

In the last years, recombinant exosomes from the hyperthermophilic Archaea *S. solfataricus*, *Archaeoglobus fulgidus* and *Pyrococcus abyssi* were analysed by crystallography, small angle X-ray scattering and mutagenesis, giving important information about the structure of this protein complex and its mechanisms.[9,10,13,18,25,26]

The Catalytically Active RPD-Hexamer

Crystallographic analyses revealed that the two RPD-containing subunits of the archaeal exosome Rrp41 and Rrp42 are arranged in a hexameric ring composed of three Rrp41/Rrp42 dimers. This hexameric ring is the minimal catalytic subunit of the complex, it exhibits phosphorolytic exoribonuclease activity in 3′ to 5′ direction and it strongly resembles the structure of the hexameric ring build of the two RPD domains present in each polypeptide of the homotrimeric bacterial PNPase.[9,10,12,20] The phosphate binding sites were visualised using a phosphate-mimicking ion and structure-guided mutations localised the active sites in Rrp41, in close vicinity to Rrp42, near the bottom of the central channel of the hexamer (Fig. 1).[9]

Structures of Rrp41/Rrp42 hexamers bound to RNA were also resolved showing that the four most distal nucleotides of an RNA substrate (N1-N4, numbering form the 3′-end) are bound in a cleft of an Rrp41-Rrp42 dimer by ionic interactions with a ladder of arginine residues.[27] Arginine side chains contributed by both Rrp41 and Rrp42 are involved in these interactions, explaining the importance of complex formation for enzymatic activity. Mutations of these arginines to glutamates abolish RNA degradation. The substrate

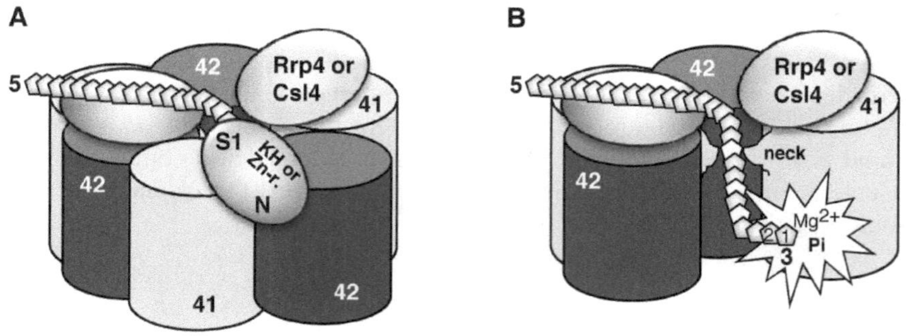

Figure 1. Schematic representation of the structure of the archaeal exosome with bound RNA. A) The nine-subunit exosome with the protruding 5′-end of an RNA substrate. Alternating RPD-containing subunits Rrp41 and Rrp42 (barrels marked with 41 and 42, respectively) are arranged in a hexameric ring. On the top of the ring, three RNA binding subunits (ovals representing Rrp4 or Csl4) are located. Rrp4 contains the N-terminal domain (N), the S1 domain and the KH domain; Csl4 contains the N-terminal domain, the S1 domain and the Zn-ribbon domain (Zn-r.). RNA (shown as a chain of pentoses) is bound to one of the Rrp4 or Csl4 subunits and its single stranded 3′-end is threaded through the S1-pore into the central channel of the hexamer. B) An Rrp41 subunit, an Rrp42 subunit and an RNA-binding subunit are removed to allow a view into the central channel of the hexamer. A narrow constriction (neck) formed by loops of the Rrp41 subunits interacts with the tenth nucleotide as numbered from the 3′-end of the substrate. These interactions as well as the interactions of the seventh and the fifth nucleotide with more than one Rrp41-Rrp42 dimer are important for RNA degradation. The active sites are located in the Rrp41 subunits. Inorganic phosphate (Pi), the first and the second nucleotide are bound directly at one of the active sites (asterisk) and the phosphoester bond between them is cleaved phosphorolytically. Most probably, the phosphate is activated by a magnesium ion also bound at the active site (for references see the text).

binding is performed mainly by electrostatic interactions with the phosphate groups of the ribose-phosphate backbone, ensuring sequence-unspecific RNA degradation. The exosome specificity towards RNA is explained by interactions with 2′-OH of the sugar.[27]

The two most distal nucleotides of the substrate (N1 and N2) are bound directly at the active site (Fig. 1) and the bond between them is cleaved phosphorolytically, releasing a ribonucleotide 5′-diphosphate (rNDP). The active site of the archaeal Rrp41 is formed by two conserved arginine residues and a catalytically active aspartate residue (D182 in *S. solfataricus* Rrp41). It was proposed that the positively charged arginine residues counteract the close positioning of two negatively charged groups (the phosphate moiety of N1 and the phosphate ion which attacks the phosphorester linkage at the 3′-terminus) and that they stabilize the transition state.[27] The key residues of the active site are conserved in archaeal exosomes and in bacterial RNase PH and PNPase.[25,27]

However, the proposed general acid/base mechanism cannot fully explain the magnesium ion dependence of RNA degradation by all these enzymes.[14,28] Recently, manganese was identified at the active site of *E. coli* PNPase.[28] Mn^{2+} can substitute for Mg^{2+} and supports catalysis, but is easier identifiable in crystal structures. The metal ion was found to be coordinated by two conserved aspartate residues and a conserved lysine residue and it was proposed that the aspartate residues support catalysis. It was also suggested that the activation of the phosphate for nucleophilic attack on the terminal phosphoester bond is metal dependent and that metal-assisted catalysis is conserved among phosphorolytic RNases including the archaeal exosome (Fig. 1).[28]

The arising rNDP product is not released simply by diffusion, but a conformational change[25] at the N1-binding site leads to controlled (active) release of the product through a conserved side channel[25,27] in concert with the entry of an inorganic phosphate ion. Then, the RNA substrate is translocated in a way that the most 3′-end nucleotide is positioned at the N1-binding site.[25] In a reaction reverse to phosphorolysis, the Rrp41-Rrp42 hexamer polymerises RNA,[9] and the polymerisation seems to follow the steps described above in the opposite direction. Degradation and polymerization (tailing) of RNA are most probably two physiologically important functions of the exosome in Archaea, similar to the functions of PNPase in bacteria and chloroplasts.[11]

Following the path of RNA/protein interaction from an active site at the bottom to the top of the hexamer, the next nucleotide, N5, interacts with two Rrp41-Rrp42 dimers, N7 also interacts with two dimers and N10 interacts with all three dimers at the neck structure near the top of the hexamer (Fig. 1).[25] These interactions are essential for RNA degradation.[25,27] The interaction of N7 and N10 with the exosome is sequence-unspecific but is mediated by the base and not by the phosphate-ribose backbone. The N7 and N10 binding sites are flexible structures of the exosome, which are stabilised upon RNA binding.[25,27]

The neck of 8 to 10 Å in diameter is formed by loops of the Rrp41 subunits and ensures that only single stranded RNA can be threaded into the channel to reach one of the active sites near the bottom of the hexamer.[27] This was demonstrated using RNA oligoribonucleotides containing a stable stem-loop structure at the 5′-end followed by poly(A) tails of different length in degradation assays with the hexameric ring of the *S. solfataricus* exosome. Only substrates carrying tails of at least 10 nt were degraded and a tail of approximately 9 nt remained intact, verifying experimentally that the single stranded RNA substrate follows a path from the top to the bottom of the hexamer, even in the absence of the RNA-binding proteins Rrp4 and Csl4.[27] This can be explained by the electrostatic surface of the hexamer, which is negatively charged at the bottom and on the side (excluding interactions with RNA in these regions) and is positively charged at the entry pore and at the central channel down to the active sites.[10]

The Flexible RNA-Binding Cap

In the nine-subunit form of the archaeal exosome, three polypeptides which can be represented by Rrp41 and/or Csl4, bind on the top of the hexameric ring forming a flat, trimeric cap with a central pore.[10] The trimeric cap (also called the RNA binding ring) contains multiple domains with RNA binding capability: Rrp4 comprises an N-terminal domain and the RNA binding S1 (ribosomal protein S1 homology) and KH (protein K homology) domains, while Csl4 is composed of an N-terminal domain and the RNA binding S1 and Zn-ribbon domains (Fig. 1).[7] Crystallographic studies revealed that the hydrophobic surfaces of the nine domains of the trimeric cap are involved in protein-protein interactions between the individual domains and between the cap and the hexamer. The remaining cap surface represents the top of the nine-subunit exosome and is suitable for interaction with RNA substrates and accessory protein factors.[10]

A comparison of the crystal structures of the two isoforms of the *A. fulgidus* exosome, containing either Rrp4 or Csl4 revealed that in both cases the S1-domains are located at the centre and form a pore which is 15 Å wide (in the case of Rrp4) or 18 Å wide (in the case of Csl4). Although Rrp4 and Csl4 are anchored to the hexameric RPD-ring via their N-terminal domains and their S1-domains form the central entry pore for the substrate, the positions of the KH and Zn-ribbon domains differ significantly in the two different isoforms. However, their positions do not exclude the existence of heterotrimeric caps, since the individual polypeptides of a cap bind independently to the RPD-hexamer and do not interact with each other. Indeed, it was possible to reconstitute recombinant exosomes with heterotrimeric caps.[10]

The spatial structure of the Rrp4-containing nine-subunit exosome of *S. solfataricus* was determined by two different groups.[18,26] Lorentzen et al resolved the structure of a symmetric protein complex and noticed that the S1 and KH domains interact closely and seem to form a single structural unit. The S1/KH unit possesses higher temperature factors than the well ordered N-terminal domain, indicating that the S1/KH part of Rrp4 is flexible.[18] In the structure resolved by Lu et al, the internal symmetry of the Rrp4-ring was broken by rigid body and thermal motions, although the intermolecular interactions between the Rrp4 ring and the RPD-hexamer of Rrp41 and Rrp42 were similar to the previously described.[26] Each of the Rrp4 subunits was found to possess distinct thermal and conformational characteristics, while the RPD-hexamer was rigid. The major difference to the structure published by Lorenzen et al is the different position of the S1 and KH domains of one of the Rrp4 subunits—these domains are moved away from the central pore, which becomes wider.[26]

These data strongly suggest that the RNA binding ring on the top of the hexamer is a highly flexible structure and several studies support this suggestion. As discussed by Lu et al,[26] major conformational differences between the exosomes of *A. fulgidus* and *S. solfataricus* are observed in the RNA-binding ring. Moreover, the S1 and KH domains of bacterial PNPases are disordered and not visible in the crystal structures, consistent with a high flexibility of these domains.[20,29] Most importantly, the analysis of the *Pyrococcus* exosome in solution by small angle X-ray scattering revealed that the Rrp4 subunits are attached to the hexameric core (presumably by the N-terminal domain) as extended and flexible arms[13] (which probably consist of the S1 and KH domains).

In addition to providing a substrate binding surface, the flexible RNA binding cap also influences the structure of the RPD-hexamer. The shape and the size of the central channel of the hexamer is somewhat different between the two isoforms of the *A. fulgidus* exosome due to differences in the Rrp41 structure.[10] The central channel of the RPD-hexamer of *S. solfataricus* is narrowed in presence of Rrp4, similarly to what was observed for *E. coli* PNPase.[29] Such structural plasticity indicates that the two different RNA binding subunits may allosterically regulate the catalytically active core of the RPD-ring and is compatible with the observation that different isoforms of the exosome differ in their activities.[10,12]

It is important to notice that Rrp4 and Csl4 are conserved in the exosomes of Archaea and Eukarya,[22,30-32] suggesting important differential roles for these proteins, most probably in substrate selection. The two different isoforms of the archaeal exosome (the Rrp4-exosome and the Csl4-exosome) harbor different RNA binding domains[7] and different electrostatic surfaces,[10] consistent with the idea that Rrp4 and Csl4 are responsible for the interaction with different molecules. Indeed, we found that the *S. solfataricus* Rrp4-exosome strongly prefers poly(A), while the Csl4-exosome more efficiently degrades heteropolymeric RNA (Roppelt, V., Klug, G., Evguenieva-Hackenberg, E., submitted).

It is assumed that RNA is bound by the S1-subunits Rrp4 and/or Csl4 and the single stranded 3′-end is threaded through the central channel to reach an active site on the bottom of the hexameric ring (Fig. 1).[18] Although the hexameric ring is sufficient for RNA degradation, the S1-subunits Rrp4 and Csl4 strongly increase RNA binding and

RNA degradation by the exosome.[10,12,13,24] Importantly, highly structured, natural tRNA cannot be degraded by the *S. solfataricus* hexamer, but is easily degraded by the Rrp4- and Csl4 exosomes.[14] Most probably, binding of tRNA by the multiple RNA binding domains at the top surface of the exosome leads to ATP-independent unwinding of secondary structures, a mechanism which was also proposed for degradation of structured RNA by RNase R and the eukaryotic exosome.[33]

PHYSIOLOGICAL FUNCTIONS

The recombinant archaeal exosome is capable to degrade RNA phosphorolytically and to perform the opposite reaction, untemplated synthesis RNA using rNDPs.[9-12,14] These activities are also exhibited by native coimmunoprecipitated exosomes and it was shown that the exoribonucleolytic and the polynucleotidylation activities of the cell-free extract of *S. solfataricus* can be assigned to the exosome.[11,12] Both activities seem to be physiologically relevant, like in the case of PNPases in bacteria and chloroplasts, which acts in vivo as exoribonucleases and as RNA-tailing enzymes.[15,16] RNA tails serve as loading platforms for exoribonucleases and tailed RNAs are thus destabilised and degraded faster than nontailed molecules.[34] Like PNPase, the archaeal exosome synthesizes heteropolymeric RNA-tails.[11,14] Sequences of posttranscriptionally added tails from exosome-containing Archaea were determined and tailed RNAs were identified as truncated mRNA and rRNA molecules. A tailed precursor of 16S rRNA was also detected.[11,35] Since RNA tailing is an event preceding and enhancing RNA decay, these data suggest a function of the exosome in targeting mRNA and rRNA fragments for degradation and in rRNA maturation.[34]

The regulation of the dual function of PNPase and the archaeal exosome is still not clear. It was proposed that local changes in the concentration of inorganic phosphate, rNDPs and Mg^{2+} contribute to this regulation, since such changes modulate the activity in vitro.[14] The reversible phosphorolytic activity probably allows to save energy avoiding the necessity to use rNTPs for synthesis of RNA tails, but the difficulty to regulate the two directions of the reaction is possibly the reason why the eukaryotic nine-subunit exosome has lost its activity and the RNA degradation and polynucleotidylation functions were separated in higher organisms.[36] The eukaryotic nine-subunit exosome is responsible for recruitment of substrates, their unwinding and channelling through the central hole of the hexameric ring, but RNA degradation is performed hydrolytically by Rrp44 which interacts with the bottom of the hexamer.[33,36-38] The untemplated synthesis of short, destabilizing poly(A)-tails to RNA is performed by different protein complexes named TRAMP in eukaryotic cells.[39,40]

Although important and probably essential aspects in RNA processing and degradation depend on the exosome, not all Archaea harbour this protein complex: in most methanogens and in halophilic Archaea, the genes encoding Rrp4, Rrp41, Rrp42 and Csl4 were lost.[7,11,35] In such organisms, posttranscriptionally added RNA-tails are not detectable.[11,35] The correlation of the presence of the exosome with the presence of heteropolymeric RNA tails in Archaea and the lack of genes encoding other polynucleotidylating enzymes in archaeal genomes supports the view that RNA tailing is a major function of the exosome. Interestingly, the exosome-less Archaea still harbour DnaG, suggesting an extraordinarily important role for this protein in RNA metabolism.[34]

CONCLUSION

Although the archaeal exosome shows structural similarity to its eukaryotic counterpart, it is functionally similar to bacterial PNPase. This is in agreement with the structure of mRNA in the third domain of life, which is similar to bacterial and not to eukaryotic mRNA—archaeal mRNA is often polycistronic, does not carry long stabilizing poly(A) tails but short, presumably destabilizing heteropolymeric tails at the 3′-end and is not capped by methylguanosine.[34] Recently we found that the *S. solfataricus* exosome is localized at the cell periphery and cosediments with membranes in sucrose gradients (Roppelt et al, submitted). The localization of the archaeal exosome at the membrane is an additional parallel between the machineries for RNA processing and degradation in Archaea and Bacteria. RNA-degrading protein complexes in *E. coli* and *Bacillus subtilis* were also shown to be membrane-bound,[41-43] suggesting a need of prokaryotic cells to spatially organize RNA processing and degradation.

ACKNOWLEDGEMENTS

I thank Gabriele Klug for generous support and helpful discussions. The work on the exosome of *S. solfataricus* was supported by the Deutsche Forschungsgemeinschaft.

NOTE ADDED IN PROOF

The two papers by Roppelt V et al were in press at the publication of this book: (1) Roppelt V, Klug G, Evguenieva-Hackenberg E. The evolutionarily conserved subunits Rrp4 and Csl4 confer different substrate specificities to the archaeal exosome. FEBS Lett. 2010 May 17. [Epub ahead of print]; and (2) Roppelt V, Hobel CF, Albers SV et al. The archaeal exosome localizes to the membrane. FEBS Lett. 2010 May 17. [Epub ahead of print].

REFERENCES

1. Woese CR, Fox GE. Phylogenetic structure of the prokaryotic domain: the primary kingdoms. Proc Natl Acad Sci USA 1977; 74:5088-5090.
2. Woese CR, Kandler O, Wheelis ML. Towards a natural system of organisms:proposal for the domains Archaea, Bacteria and Eucarya. Proc Natl Acad Sci USA 1990; 87:4576-4579.
3. Kletzin A. General characteristics and important model organisms. In: Cavicchioli R., ed. Archaea—Molecular and Cellular Biology, Washington, DC: ASM Press 2007:14-92.
4. Barry ER, Bell SD. DNA replication in the archaea. Microbiol Mol Biol Rev 2006; 70:876-887.
5. Werner F, Weinzierl RO. A recombinant RNA polymerase II-like enzyme capable of promoter-specific transcription. Mol Cell 2002; 10:635-646.
6. Anantharaman V, Koonin EV, Aravind L. Comparative genomics and evolution of proteins involved in RNA metabolism. Nucleic Acids Res 2002; 30:1427-1464.
7. Koonin EV, Wolf YI, Aravind L. Prediction of the archaeal exosome and its connections with the proteasome and the translation and transcription machineries by a comparative-genomic approach. Genome Res 2001; 11:240-252.
8. Evguenieva-Hackenberg E, Walter P, Hochleitner E et al. An exosome-like complex in Sulfolobus solfataricus. EMBO Rep 2003; 4:889-893.

9. Lorentzen E, Walter P, Fribourg S et al. The archaeal exosome core is a hexameric ring structure with three catalytic subunits. Nat Struct Mol Biol 2005; 12:575-581.
10. Büttner K, Wenig K, Hopfner KP. Structural framework for the mechanism of archaeal exosomes in RNA processing. Mol Cell 2005; 20:461-471.
11. Portnoy V, Evguenieva-Hackenberg E, Klein F et al. RNA polyadenylation in Archaea: not observed in Haloferax while the exosome polynucleotidylates RNA in Sulfolobus. EMBO Rep 2005; 6:1188-1193.
12. Walter P, Klein F, Lorentzen E et al. Characterization of native and reconstituted exosome complexes from the hyperthermophilic archaeon Sulfolobus solfataricus. Mol Microbiol 2006; 62:1076-1089.
13. Ramos CR, Oliveira CL, Torriani IL et al. The Pyrococcus exosome complex: structural and functional characterization. J Biol Chem 2006; 281:6751-6759.
14. Evguenieva-Hackenberg E, Roppelt V, Finsterseifer P et al. Rrp4 and Csl4 are needed for efficient degradation but not for polyadenylation of synthetic and natural RNA by the archaeal exosome. Biochemistry 2008; 47:13158-13168.
15. Mohanty BK, Kushner SR. Polynucleotide phosphorylase functions both as a 3′ right-arrow 5′ exonuclease and a poly(A) polymerase in Escherichia coli. Proc Natl Acad Sci USA 2000; 97:11966-11971.
16. Rott R, Zipor G, Portnoy V et al. RNA polyadenylation and degradation in cyanobacteria are similar to the chloroplast but different from Escherichia coli. J Biol Chem 2003; 278:15771-15777.
17. Farhoud MH, Wessels HJ, Steenbakkers PJ et al. Protein complexes in the archaeon Methanothermobacter thermautotrophicus analyzed by blue native/SDS-PAGE and mass spectrometry. Mol Cell Proteomics 2005; 4:1653-1663.
18. Lorentzen E, Dziembowski A, Lindner D et al. RNA channelling by the archaeal exosome. EMBO Rep 2007; 8:470-476.
19. Liu Q, Greimann JC, Lima CD. Reconstitution, activities and structure of the eukaryotic RNA exosome. Cell 2006; 127:1223-1237. Erratum in: Cell 2007; 131:188-189.
20. Symmons MF, Jones GH, Luisi BF. A duplicated fold is the structural basis for polynucleotide phosphorylase catalytic activity, processivity and regulation. Structure 2000; 8:1215-1226.
21. Walter P. Aufklärung von Struktur und Funktion des archaealen Exosoms durch Charakterisierung nativer und rekonstituierter Proteinkomplexe aus Sulfolobus solfataricus. Dissertation, 2007: Universität Giessen, Germany.
22. Mitchell P, Petfalski E, Shevchenko A et al. The exosome: a conserved eukaryotic RNA processing complex containing multiple 3′—>5′ exoribonucleases. Cell 1997; 91:457-466.
23. Chekanova JA, Dutko JA, Mian IS et al. Arabidopsis thaliana exosome subunit AtRrp4p is a hydrolytic 3′—>5′ exonuclease containing S1 and KH RNA-binding domains. Nucleic Acids Res 2002; 30:695-700.
24. Oddone A, Lorentzen E, Basquin J et al. Structural and biochemical characterization of the yeast exosome component Rrp40. EMBO Rep 2007; 8:63-69.
25. Navarro MV, Oliveira CC, Zanchin NI et al. Insights into the mechanism of progressive RNA degradation by the archaeal exosome. J Biol Chem 2008; 283:14120-14131.
26. Lu C, Ding F, Ke A. Crystal structure of the S. solfataricus archaeal exosome reveals conformational flexibility in the RNA-binding ring. PLoS One 2010; 5:e8739.
27. Lorentzen E, Conti E. Structural basis of 3′ end RNA recognition and exoribonucleolytic cleavage by an exosome RNase PH core. Mol Cell 2005; 20:473-481.
28. Nurmohamed S, Vaidialingam B, Callaghan AJ et al. Crystal structure of Escherichia coli polynucleotide phosphorylase core bound to RNase E, RNA and manganese: implications for catalytic mechanism and RNA degradosome assembly. J Mol Biol 2009; 389:17-33.
29. Shi Z, Yang WZ, Lin-Chao S et al. Crystal structure of Escherichia coli PNPase: central channel residues are involved in processive RNA degradation. RNA 2008; 14:2361-2371.
30. van Hoof A, Staples RR, Baker RE et al. Function of the ski4p (Csl4p) and Ski7p proteins in 3′-to-5′ degradation of mRNA. Mol Cell Biol 2000; 20:8230-8243.
31. Raijmakers R, Noordman YE, van Venrooij WJ et al. Protein-protein interactions of hCsl4p with other human exosome subunits. J Mol Biol 2002; 315:809-818.
32. Haile S, Estevez AM, Clayton C. A role for the exosome in the in vivo degradation of unstable mRNAs. RNA 2003; 9:1491-1501.
33. Bonneau F, Basquin J, Ebert J et al. The yeast exosome functions as a macromolecular cage to channel RNA substrates for degradation. Cell 2009; 139(3):547-59.
34. Evguenieva-Hackenberg E, Klug G. RNA degradation in Archaea and Gram-negative bacteria different from Escherichia coli. Prog Mol Biol Transl Sci 2009; 85:275-317.
35. Portnoy V, Schuster G. RNA polyadenylation and degradation in different Archaea; roles of the exosome and RNase R. Nucleic Acids Res 2006; 34:5923-5931.
36. Hartung S, Hopfner KP. Lessons from structural and biochemical studies on the archaeal exosome. Biochem Soc Trans 2009; 37:83-87.

37. Dziembowski A, Lorentzen E, Conti E et al. A single subunit, Dis3, is essentially responsible for yeast exosome core activity. Nat Struct Mol Biol 2007; 14:15-22.
38. Wang HW, Wang J, Ding F et al. Architecture of the yeast Rrp44 exosome complex suggests routes of RNA recruitment for 3′ end processing. Proc Natl Acad Sci USA 2007; 104:16844-16849.
39. LaCava J, Houseley J, Saveanu C et al. RNA degradation by the exosome is promoted by a nuclear polyadenylation complex. Cell 2005; 121:713-724.
40. Vanácová S, Wolf J, Martin G et al. A new yeast poly(A) polymerase complex involved in RNA quality control. PLoS Biol 2005; 3:e189.
41. Liou GG, Jane WN, Cohen SN et al. RNA degradosomes exist in vivo in Escherichia coli as multicomponent complexes associated with the cytoplasmic membrane via the N-terminal region of ribonuclease E. Proc Natl Acad Sci USA 2001; 98:63-68.
42. Taghbalout A, Rothfield L. RNase E and RNA helicase B play central roles in the cytosceletal organization of the RNA degradosome. J Biol Chem 2008; 283:13850-13855.
43. Commichau FM, Rothe FM, Herzberg C et al. Novel activities of glycolytic enzymes in Bacillus subtilis: interactions with essential proteins involved in mRNA processing. Mol Cell Proteomics 2009; 8:1350-1360.

CHAPTER 4

THE EXOSOMES OF TRYPANOSOMES AND OTHER PROTISTS

Christine Clayton*[,1] and Antonio Estevez[2]

[1]*Zentrum für Molekulare Biologie der Universität Heidelberg, DKFZ-ZMBH Alliance, Heidelberg, Germany;*
[2]*Instituto de parasitologia y Biomedicina "Lopez-Neyra", CSIC, Avda. del Conocimiento, Granada, Spain.*
**Corresponding Author: Christine Clayton—Email: cclayton@zmbh.uni-heidelberg.de*

Abstract: The archaeal exosome contains three heterodimeric RNase PH subunits, forming a hexamer with RNase activity; on top sits a trimer of two different SI domain proteins. In animals and yeast, six different, but related subunits form the RNase PH-like core, but these lack enzyme activity; there are three different SI-domain proteins and enzyme activity is provided by the endo/exonuclease Rrp44 or—mainly in the nuclear exosome—the Rnase D enzyme Rrp6. Trypanosomes diverged from yeast and mammals very early in eukaryotic evolution. The trypanosome exosome is similar in subunit composition to the human exosome, but instead of being an optional component, trypanosome RRP6 is present in the nucleus and cytoplasm and is required for exosome stability. As in human cells and yeast, the trypanosome exosome has been shown to be required for processing and quality control of rRNA and to be involved in mRNA degradation. Electron microscopy results for a *Leishmania* exosome suggest that RRP6 is located on the side of the RnasePH ring, interacting with several exosome core proteins. Results of a search for exosome subunits in the genomes of widely diverged protists revealed varied exosome complexity; the *Giardia* exosome may be as simple as that of Archaea.

INTRODUCTION

Most studies of eukaryotic exosomes so far have focussed either on animals and yeast, or on Archaea. The archaeal exosome contains three heterodimeric RNase PH subunits, forming a hexamer; on top sits a trimer of two different SI domain proteins.[1] In animals and yeast, six different, but related subunits form the RNase PH core and there are also three different SI-domain proteins.[2] In addition, the yeast exosome contains the

RNA Exosome, edited by Torben Heick Jensen.

exo/endoribonuclease Rrp44 and both yeast and human exosomes can associate with an RNase D enzyme, Rrp6[3] and an additional protein, Rrp47.[4] Although plant exosomes appear to be similar to those of yeast,[5-7] we do not know if this can be generalised to all other eukaryotes. A eukaryotic exosome similar to that of Archaea can readily be envisaged. If exosomes of intermediate complexity were to be discovered, studies of their function could yield valuable insights into possible roles of the different subunits in the human exosome and into the reasons for sequence divergence in the core subunits. Sequence comparisons from highly diverged eukaryotes can also pinpoint highly conserved residues that are likely to be essential for exosome function.

The most important properties of the exosome are its catalytic activity and its regulation. Both depend on the properties of the core exosome subunits and their interaction with other factors. Even the enzymatic activity, however, is not thoroughly understood. The core RNase PH subunits in yeast and human exosomes have lost catalytic activity,[8] but *Arabidopsis* RRP41 retains the catalytic site and was shown to be active.[9] How many other eukaryotic species retain enzyme activity in the RNase PH subunits? In yeast, catalysis depends on the activities of two associated proteins, Rrp44 and Rrp6.[10-12] In humans, some exosomes may depend on Rrp6 for activity, but there is only weak evidence for an association with Rrp44.[13] Are Rrp44 and/or Rrp6 always present in eukaryotic cells? If so, are they exosome-associated? Can Rrp44 and Rrp6 act alone and to what extent are their activities guided by exosome association? Which exosome functions—and which modes of regulation—are conserved in all eukaryotic cells and which are specialised for particular species? Information from a wider range of species may help us to better answer some of these questions and could clarify the pathway by which the Archaeal complex increased in complexity.

There are six major groups of eukaryotes, which separated soon after the acquisition of the mitochondrion. Animals and yeast are both in one group, the *Opisthokonta*. *Plantae* are a second group. In this chapter we describe, in detail, the properties of the exosome from a representative of a third group and also briefly examine evidence for the exosome in other deeply divergent eukaryotes.

TRYPANOSOMES AS A MODEL SYSTEM FOR RNA PROCESSING AND TURNOVER

The Kinetoplastids, which include trypanosomes and leishmanias, are protists and belong to the *Excavata*, which branched away from the *Plantae* and *Opisthokonta* extremely early in eukaryotic evolution. Trypanosomes and leishmanias are parasites of diverse animals and plants and cause several diseases of medical, veterinary or agricultural importance, particularly in developing countries. For this reason alone, they have been studied with the aim of finding new chemotherapies or vaccines. However there is an additional motivation for studying trypanosome exosomes: trypanosomes are an excellent model system for studies of RNA processing and decay. In kinetoplastid organisms, transcription of protein-coding genes by RNA polymerase II is polycistronic and appears to be constitutive;[14] individual mRNAs are generated by 5′ *trans* splicing and 3′ polyadenylation.[15] The polycistronic units contain many mRNAs with unrelated functions and regulation patterns,[16,17] so control of gene expression is posttranscriptional. Most evidence so far has pinpointed mRNA degradation and translation as the major control points.[18,19] One consequence of this is that investigations of mRNA degradation and processing are not complicated by transcriptional effects: this greatly simplifies quantitative analysis and interpretation of experimental results. A second consequence is that trypanosomes are uniquely reliant upon mRNA degradation for their survival: thus the enzymes that digest RNA and the proteins that regulate them, may constitute promising targets for new anti-trypanosomal drugs.

The African trypanosome *Trypanosoma brucei* is a particularly convenient subject for molecular investigation because it is readily culturable and reverse genetics, including convenient RNAi methods, are well established.

THE STRUCTURE OF THE TRYPANOSOME EXOSOME

The trypanosome exosome was purified by tandem affinity chromatography, using TAP-tagged *Tb*RRP4 as the ligand. Mass spectrometry revealed the presence of three different S1 subunits, *Tb*RRP4, *Tb*RRP40 and *Tb*CSL4 and six RNase PH subunits, which were designated *Tb*RRP41A, *Tb*RRP41B, *Tb*RRP45, *Tb*EAP1, *Tb*EAP2 and *Tb*EAP4.[20,21] (The "EAP" stands for "Exosome Associated Protein"; although these proteins have RNase PH-like domains, they could not be assigned as homologues of any particular yeast subunit on the basis of sequence alignment alone.) In addition, the complex contained the RNase D-like protein *Tb*RRP6 and *Tb*EAP3—a likely homologue of yeast Rrp47. A gene encoding *Tb*RRP44 was found and the expression of the protein was demonstrated. *Tb*RRP44 was however not present in the purified exosomes and did not sediment with the exosome on glycerol gradients.[21]

To determine the relative arrangements of the different subunits, each subunit was depleted using RNA interference and the effects on the abundances of the other subunits and the migration of the exosome on glycerol gradients, were monitored.[20] The complex seemed intact after depletion of any of the S1 subunits. In contrast, depletion of the RNase PH domain proteins *Tb*RRP41A and *Tb*EAP1 caused exosome disassembly, as judged by the distribution of *Tb*RRP45 on the glycerol gradient. Depletion of any of the core PH subunits caused decreases in the abundance of *Tb*RRP4. The results suggested that some of the RNase PH and S1 subunits were unstable if they were not associated in the complex. Strikingly, depletion of *Tb*RRP6 and *Tb*EAP3 caused a loss of both *Tb*RRP45 and *Tb*RRP4, suggesting that—in contrast to the situation in yeast or human cells[22]—*Tb*RRP6 might have a structural role in the trypanosome exosome. Depletion of *Tb*RRP44 had no effect on the exosome.

The mutual interactions of the subunits were also tested by yeast two-hybrid analysis.[20] The results suggested an overall arrangement of the six RNase PH subunits that was similar to that seen in yeast and indicated mutual links between the S1 domain proteins as well as binding of these to the PH core (Fig. 1A). They also showed an interaction between RRP6 and EAP3, but neither these two proteins, nor RRP44, showed interactions with the core subunits.

THE POSITION OF RRP6 RELATIVE TO THE EXOSOME CORE

The overall picture so far was of an eleven-subunit complex that contained RRP6 but not RRP44. The fact that RRP6 depletion affected the rest of the exosome was surprising, given that this protein is associated only with the nuclear exosome in yeast. The results of localisation studies[23] indicated that in contrast, *Tb*RRP6 is in

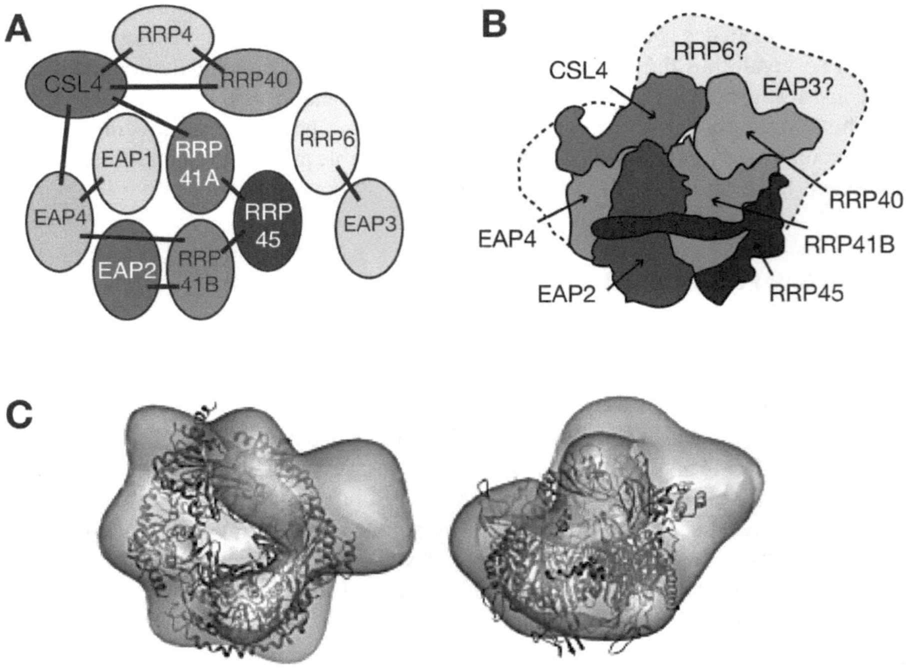

Figure 1. A) Interactions of trypanosome exosome subunits, detected using the yeast 2-hybrid system. Colour coding is the same as in (B). B) Diagrammatic representation of our space-filling model of the kinetoplastid exosome, labelled with the names of the trypanosome proteins, taken from ref. 24. The blue proteins are the RNase PH subunits and the red-pink ones are S1 subunits. Dotted lines represent the outline of the *Leishmania tarentolae* exosome as determined by electron microscopy; the yellow portion is postulated to contain RRP6 and EAP3. C) Approximate fit of the structure of the nine subunits of the human exosome[2] into the EM map, taken from ref. taken from ref. 24. The grey is the *Leishmania* EM structure and the ribbon diagrams are the human exosome, with PH subunits in blue and S1 subunits in red. The image on the left is viewed from above the S1 subunits and the image on the right is viewed from the side. B,C) Used with permission from Cristodero et al, Mol Biochem Parasitol 2008; 159(1):24-9.[24] Copyright 2008, Elsevier. A color version of this image is available at www.landesbioscience.com/curie.

the trypanosome cytoplasm as well as nucleus. In addition, *Tb*RRP6 was found in a cytosolic exosome preparation.[21] *Tb*RRP6 was not affected by depletion of core exosome subunits and could also be over-expressed, so its stability does not depend upon exosome association.[23] Nevertheless, in the absence of over-expression, all *Tb*RRP6 cosedimented with the exosome, suggesting that in vivo, its expression is coregulated with that of other exosome subunits.

All published crystal structures of eukaryotic exosomes describe complexes that have been reconstituted from subunits produced in *E. coli*. The reliance of trypanosomes (and related parasites) on mRNA degradation appears to have resulted in an unusually large amount of exosome per cell. Using the parasite *Leishmania tarentolae*, which grows to high densities in an economical (serum-free) medium it was therefore possible to prepare sufficient pure native exosome for structural analysis. As a consequence, the first structure of an exosome containing homologues of Rrp6 and Rrp47 could be determined by electron microscopy.[24] In addition to the expected ring structure, which could be aligned with that of the yeast exosome, the *Leishmania tarentolae* exosome exhibited a prominent bulge on one side: presumably this represented *Tb*RRP6 and *Tb*EAP3 (Fig. 1B). The fact that *Tb*RRP6 appeared to contact several core subunits, including *Tb*RRP41A and *Tb*RRP40, might explain why *Tb*RRP6 is required for trypanosome exosome stability. This position of *Tb*RRP6 appeared to preclude "feeding" of RNAs to the protein via the central RNase PH channel. If yeast Rrp6 occupies a similar position, this would allow simultaneous binding of Rrp44 to the "bottom" of the core. If it proves difficult to obtain crystals of reconstituted Rrp6-containing exosomes, it would be worthwhile to attempt further optimisation of the *Leishmania* preparations for higher-resolution structural studies.

It was initially reported that purified exosome and recombinant RRP4 had 3′-5′ exoribonuclease activity.[21] This was however most likely due to contamination, as seen for other eukaryotic exosomes.[8] Multiple tests with the purified *Leishmania* exosome failed to detect exoribonuclease activity, despite the presence of RRP6.[24] This could be a technical artefact, or could indicate that additional cofactors or specific substrates are required for trypanosome exosome activity. Recombinant yeast Rrp6 activity appears not to be affected by association with the exosome;[2,8] native recombinant trypanosome *Tb*RRP6 is not available for testing, but the catalytic residues are conserved.

FUNCTIONS OF THE TRYPANOSOME EXOSOME IN THE NUCLEUS

The functions of the trypanosome exosome were tested by examining various RNAs after depletion of individual subunits by RNA interference, using stable cell lines in which expression of dsRNA was induced using a tetracycline-inducible promoter. In trypanosomes, the efficiency of RNA interference not only varies between cell lines, but is also influenced by cultivation and freezing of cell lines. Also, since antisera are not available for most subunits, the extent of protein depletion usually could not be monitored. It is therefore not possible to make quantitative judgements about exactly how much each protein was required for a particular process. The results of the RNAi nevertheless suggested that all exosome subunits and *Tb*RRP44 were either essential for trypanosome survival, or at least required for normal growth.

The first function of the nuclear exosome that was tested was the trimming of 5.8S rRNA precursors. Exosome subunit depletion resulted in accumulation of a 7S precursor and some intermediates, with a decrease in a 6S intermediate.[21] The most severe defects were seen with RNAi targeting *Tb*RRP4 and *Tb*RRP45, although this could have been a reflection of RNAi efficiency. Notably, despite the lack of any evidence for association of *Tb*RRP44 with the exosome, depletion of *Tb*RRP44 was equally effective in inhibiting 5.8S rRNA maturation.[21] Clearly, therefore, 5.8S rRNA maturation requires both the core exosome and *Tb*RRP44, despite the lack of any evidence for a physical association between them. Possibly, the interaction is weak or transient, or substrate-dependent. It was notable that despite precursor accumulation, the steady-state level of 5.8S RNA was not affected; perhaps the decrease in rRNA processing was matched by the decrease in trypanosome growth.

In yeast, processing of nuclear RNAs is regulated by the TRAMP complex, which contains the helicase Mtr4, one of two possible RNA-binding proteins, Air1p or Air2p and a poly(A) polymerase, either Trf4 or Trf5. This complex catalyses the addition of oligo(A) tails to exosome targets, which primes them for exosome-mediated degradation.[25,26]

A search of the trypanosome genome revealed the presence of genes encoding trypanosome MTR4 and a nuclear poly(A) polymerase, *Tb*NPAPL. Air1 and Air2 are not sufficiently conserved, at the sequence level, to allow clear identification even in the human genome and there were no obvious homologues in trypanosomes. RNAi experiments demonstrated that *Tb*MTR4 and *Tb*NPAPL were required for cell growth, and *Tb*MTR4 depletion caused defects in 5.8S rRNA maturation that were indistinguishable from those seen after *RRP6* RNAi.[27] (It should be remembered that in trypanosomes, *RRP6* RNAi affects the entire exosome.) To find out whether polyadenylation was also involved in this, a small rRNA from the 3′ end of the precursor, SR6, and a partially processed precursor, ITS7, were examined; these were chosen because these had been shown to be polyadenylated in a previous study.[28] The presence of the polyadenylated RNAs was confirmed by RT-PCR.[27] Depletion of *Tb*MTR4 increased the lengths of the poly(A) tails, while depletion of *Tb*NPAPL caused a decreased in the overall abundance of the tails. Neither co-immunoprecipitation nor affinity purification, however, revealed any evidence for the presence of a MTR4-NPAPL complex or for physical interaction with the exosome.[27] Thus although polyadenylation probably plays a role in exosome-mediated trimming of rRNA precursors, there was no evidence for the presence of a TRAMP complex. In human cells, Mtr4 interacts with exosomal cofactor Rrp47 via another protein, MPP6; it is not yet clear whether trypanosomes have a functional equivalent of MPP6 or not.

Trypanosomes have a multitude of snoRNAs and snRNAs, which might be subject to trimming by the exosome. Preliminary results suggest that indeed, *Tb*MTR4 is involved in controlling levels of snRNAs and some, but not all, snoRNAs (S. Michaeli, Bar-Ilan University, Israel, personal communication). Other possible roles for the exosome might be in the destruction of fragments generated by the RNAi machinery, either in the nucleus or cytoplasm; this has yet to be tested. The full extent of trypanosome exosome function will be revealed when high-throughput sequencing data becomes available for subunit-depleted organisms.

FUNCTIONS OF THE TRYPANOSOME EXOSOME IN THE CYTOPLASM

African trypanosomes replicate in the gut of Tsetse flies and in body fluids of a mammalian host. The two replicative forms, which are called “procyclic forms” and “bloodstream forms” respectively, differ in their energy metabolism and in surface protein expression. The bloodstream forms express variant surface glycoprotein (VSG) and rely heavily on glycolysis, while the procyclic forms express EP procyclin and have more complex energy metabolism. Depending on the criteria adopted, between 200 and 1000 mRNAs (out of about 8000 open reading frames) change in abundance as the parasites differentiate from the bloodstream form to the procyclic form.[29-31] Regulation of mRNA turnover plays a major role in this: the half-lives of mRNAs range from several hours to minutes.[18,32] For example, mRNAs encoding mitochondrial enzymes are more stable in procyclic forms than in bloodstream forms.[33] Experiments with reporter genes have demonstrated that the half-lives of most trypanosome mRNAs tested are determined by sequences in their 3′-untranslated regions.[18,19]

About 5% of the bloodstream form mRNA encodes VSG[32] while in the procyclic form, 3% of mRNA encodes EP procyclin. These genes are, exceptionally, transcribed by RNA polymerase I[34] and subject to transcriptional regulation. *VSG* genes are completely shut off in procyclic forms, but the *EP* gene transcription is down-regulated only 10-20 fold in bloodstream-forms. The steady-state *EP* mRNA levels are however much more strongly regulated: *EP* mRNA has a half-life of hours in procyclic forms, but is degraded within minutes in bloodstream forms. The half-life is determined mainly by U-rich region in the 3′-untranslated region.[35]

In both mammalian cells and yeast, classical mRNA degradation is generally initiated by removal of the poly(A) tail. This is followed by decapping and 5′-3′ degradation and/or degradation from the 3′-end by the exosome (see e.g., ref. 36). Exceptionally, RNAs may be subject to deadenylation-independent decapping. In addition they can be subject to endonucleolytic cleavage, either by specialised endonucleases or by the RISC complex; here too the products become exosome substrates (see e.g., ref. 37).

To determine the direction of mRNA degradation in trypanosomes and leishmanias, various approaches were undertaken. Using reporter genes, secondary structures were introduced to inhibit exoribonuclease progression and degradation intermediates were mapped by RNase H digestion.[38,39] In addition, in trypanosomes, the effects of RNAi targeting the exosome,[40] deadenylases[39,41] and *Tb*XRNA (the major mRNA-targeted 5′-3′ exoribonuclease[42]) were assessed. Most of the work was done in bloodstream-form trypanosomes. The results showed that the decay of mRNAs with half-lives over 15 min was initiated by deadenylation, as usual.[41] In contrast, two very unstable mRNAs—including the EP procyclin mRNA—seemed to be subject to two pathways. One started with deadenylation,[39] while the other involved direct attack on the 5′-ends of polyadenylated mRNA.[42] For these very unstable mRNAs, there was some evidence that the exosome was involved in 3′-5′ digestion of deadenylated mRNA.[43] Exosome depletion led to a delay in the decay of reporter mRNAs with three different stabilising 3′-UTRs[40] and stabilised deadenylated RNA. It did not, paradoxically, lead to a detectable increase in the steady-state abundances of the tested RNAs. Overall, therefore, the evidence suggests that the exosome plays a relatively minor role in trypanosome mRNA degradation. Experiments to examine the pathway of mRNA degradation in the related parasite *Leishmania infantum* also provided evidence for the existence of both 5′-3′ and 3′-5′ pathways.[38]

In other organisms, proteins involved in mRNA degradation are concentrated in "processing bodies" (P-bodies).[44] After various stresses, some of these proteins, but also translation factors and polyadenylated mRNA, become concentrated in stress granules.[45] The current consensus is that the exosome is not located in either of these structures. In trypanosomes, antibody to *Tb*RRP6 detected tiny granules spread throughout the trypanosome cytoplasm as well as a more uniform staining of the nucleus;[23] the granules could however have been a fixation artefact. Inducibly expressed TAP-tagged *Tb*RRP4 was concentrated in the nucleus, while TAP-tagged *Tb*EAP1 was more uniformly spread in the cytoplasm as well. There is no unequivocal evidence for the presence of P-bodies in trypanosomes, but stress granules can be induced by heat shock[46] or starvation.[47] It is not known whether these granules contain exosome components.

The yeast Lsm complex preferentially binds to the 3′ ends of oligoadenylated mRNAs and is thought to be involved in the recruitment of the decapping complex and possibly also the exosome.[48] Exosome-dependent mRNA degradation in yeast is also stimulated by the Ski complex. Ski proteins are not present in trypanosomes, but an Lsm complex is. In yeast, the 7-subunit Lsm complex has a compartment—specific component—the Lsm2 of the nuclear complex is replaced by Lsm1 in the cytosol. The trypanosome Lsm complex consists of LSM2-LSM8 in both compartments and depletion of either *Tb*LSM3 or *Tb*LSM8 inhibited degradation of two different mRNAs.[49]

CONSERVATION OF THE EXOSOME IN EUKARYOTIC EVOLUTION

To determine the degree of exosome conservation beyond trypanosomes, plants and Opisthokonts, we have performed BLASTp analysis with a small number of highly diverged genomes. We used yeast or human exosome subunit query sequences and checked dubious matches by reciprocal BLASTp and by domain searches. In some cases we could also search directly for annotated RNase PH domains. The genomes searched were of *Dictyostelium discoideum, Plasmodium berghei*, two ciliates (*Paramecium tetraurelia, Tetrahymena thermophila*), the Chromalveolate *Thalassiosira pseudonana*; and *Trichomonas vaginalis*, *Giardia lamblia* and *Giardia intestinalis* (like trypanosomes, members of the Excavata) (Fig. 2). For all results we must bear in mind that genome assemblies may be incomplete and that bona fide homologues can be missed in BLASTp searches due to low sequence conservation.

The numbers of RNase PH subunits found was variable. Interestingly, there were usually fewer than 6 different subunits; these always, however, included one or more versions of Rrp45. The minimal set was one Rrp45 and one other RNase PH similar to Rrp41 or Rrp43. This small set is seen in the two available *Giardia* genomes (C6LR86/A8BNT9 and A8B493/C6LWS9) and in the two ciliates—for *Tetrahymena*, Q23PW9 and EAR84452. *Tetrahymena* has an additional Rrp45 which, despite being almost double the normal size, has only one RNase PH domain. The arrangement of the PH subunits in these organisms would be of interest—are there three copies of each subunit, arranged as dimers, as in Archaea?

Most of the searched organisms had two or three S1 domain subunits, but *Trichomonas* had only one. Mtr3 was not found; Rrp46 was present only in *Dictyostelium*. Rrp44 was conserved in all of the searched genomes, but we did not find any trace of Rrp6 in *Plasmodium* or *Giardia*. If Rrp6 is missing, but the exosome is active in

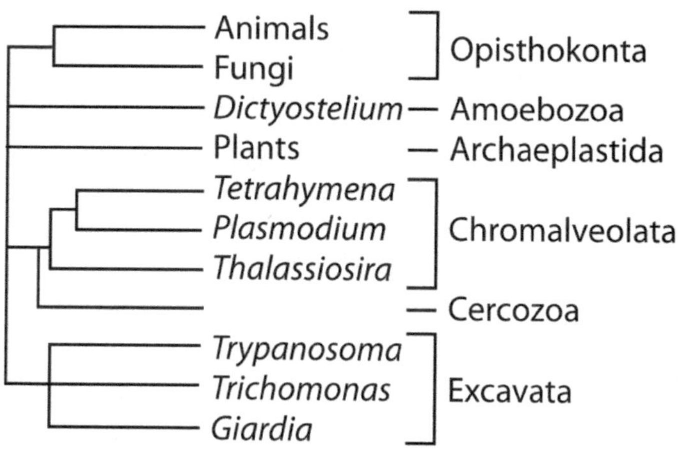

Figure 2. Relative positions of *Dictyostelium discoideum, Plasmodium berghei, Tetrahymena thermophila, Thalassiosira pseudonana, Trichomonas vaginalis* and *Giardia* on an evolutionary tree. *Paramecium tetraurelia* is a ciliate, like *Tetrahymena*. The six major groups are on the right. The diagram is heavily simplified from reference 50; none of the branch lengths are to scale. No full genome is currently available for Cercozoa. For more details see also http://tolweb.org/tree/.

these species, the 3′-5′ exoribonuclease activity could reside in the PH subunits themselves; alternatively, it might be conferred by Rrp44 should Rrp44 be exosome-associated.

This brief survey suggests that in some eukaryotes, the exosome shows a complexity that is intermediate between those found in Archaea and Opisthokonts. Since *Giardia* had two S1 subunit sequences, two RnasePHs and no Rrp6, its exosome appears to be as simple as that of Archaea. A more thorough genomic search, followed up with biochemical analysis, will be needed to find out whether this is indeed the case.

CONCLUSION

Trypanosomes are heavily reliant on posttranscriptional mechanisms for regulation of gene expression, with mRNA degradation playing a crucial role. The trypanosome exosome is, correspondingly, as complex as that of human cells. As in human cells and yeast, the trypanosome exosome has been shown to be required for processing and quality control of rRNA and to be involved in mRNA degradation. The origin of RNase activity in the trypanosome exosome remains a mystery since none was demonstrated in purified exosomes containing *Tb*RRP6. It is also not known whether *Tb*RRP44 ever associates with the trypanosome exosome: further investigation of *Tb*RRP44 would be warranted. Similarly, a thorough analysis of exosomes in widely diverged eukaryotes is likely to yield insights into exosome function. In particular, the exosome of *Giardia* appears to have Archaeal composition although it must—presumably—fulfil the functions required of the eukaryotic complex.

ACKNOWLEDGEMENTS

Figures 1B and 1C are reprinted from part of Figure 1 in Molecular and Biochemical Parasitology vol 159, pp 24-29, with permission from Elsevier. We thank Prof. Shula Michaeli (Bar-Ilan University, Israel) for communicating unpublished results.

REFERENCES

1. Lorentzen E, Dziembowskiw A, Lindner D et al. RNA channelling by the archaeal exosome. EMBO Rep 2007; 8:470-476.
2. Liu Q, Greimann JC, Lima CD. Reconstitution, activities and structure of the eukaryotic RNA exosome. Cell 2006; 127:1223-1237.
3. Raijmakers R, Schilders G, Pruijn GJ. The exosome, a molecular machine for controlled RNA degradation in both nucleus and cytoplasm. Eur J Cell Biol 2004; 83:175-183.
4. Mitchell P, Petfalski E, Houalla R et al. Rrp47p is an exosome-associated protein required for the 3′ processing of stable RNAs. Mol Cell Biol 2003; 23:6982-6992.
5. Chekanova JA, Shaw RJ, Wills MA et al. Poly(A) tail-dependent exonuclease AtRrp41p from Arabidopsis thaliana rescues 5.8 S rRNA processing and mRNA decay defects of the yeast ski6 mutant and is found in an exosome-sized complex in plant and yeast cells. J Biol Chem 2000; 275.
6. Zimmer S, Fei Z, Stern D. Genome-based analysis of Chlamydomonas reinhardtii exoribonucleases and poly(A) polymerases predicts unexpected organellar and exosomal features. Genetics 2008; 179:125-136.
7. Lange H, Holec S, Cognat V et al. Degradation of a polyadenylated rRNA maturation by-product involves one of the three RRP6-like proteins in Arabidopsis thaliana. Mol Cell Biol 2008; 28:3038-3044.

8. Liu Q, Greimann J, Lima C. Reconstitution, Activities and Structure of the Eukaryotic RNA Exosome. Cell 2007; 131:188-189.
9. Chekanova J, Dutko J, Mian I et al. Arabidopsis thaliana exosome subunit AtRrp4p is a hydrolytic 3'—>5' exonuclease containing S1 and KH RNA-binding domains. Nucleic Acids Res 2002; 30:695-700.
10. Dziembowski A, Lorentzen E, Conti E et al. A single subunit, Dis3, is essentially responsible for yeast exosome core activity. Nat Struct Mol Biol 2007; 14:15-22.
11. Schaeffer D, Tsanova B, Barbas A et al. The exosome contains domains with specific endoribonuclease, exoribonuclease and cytoplasmic mRNA decay activities. Nat Struct Mol Biol 2008; 16:56-62.
12. Schneider C, Anderson J, Tollervey D. The exosome subunit rrp44 plays a direct role in RNA substrate recognition. Mol Cell 2007; 27:324-331.
13. Lehner B, Sanderson CM. A protein interaction framework for human mRNA degradation. Genome Res 2004; 14:1315-1323.
14. Palenchar JB, Bellofatto V. Gene transcription in trypanosomes. Mol Biochem Parasitol 2006; 146:135-141.
15. Liang X, Haritan A, Uliel S et al. Trans and cis splicing in trypanosomatids: mechanism, factors and regulation. Eukaryot Cell 2003; 2:830-840.
16. Ivens ACea. The genome of the kinetoplastid parasite, Leishmania major. Science 2005; 309-442.
17. Berriman Mea The genome of the African trypanosome, Trypanosoma brucei. Science 2005; 309:416-422.
18. Clayton C, Shapira M. Post-transcriptional regulation of gene expression in trypanosomes and leishmanias. Mol Biochem Parasitol 2007; 156:93-101.
19. Haile S, Papadopoulou B. Developmental regulation of gene expression in trypanosomatid parasitic protozoa. Curr Opin Microbiol 2007; 10:569-577.
20. Estévez AM, Lehner B, Sanderson CM et al. The roles of inter-subunit interactions in exosome stability. J Biol Chem 2003; 278:34943-34951.
21. Estévez A, Kempf T, Clayton CE. The exosome of Trypanosoma brucei. EMBO J 2001; 20:3831-3839.
22. van Dijk E, Schilders G, Pruijn G. Human cell growth requires a functional cytoplasmic exosome, which is involved in various mRNA decay pathways. RNA 2007; 13:1027-1035.
23. Haile S, Cristodero M, Clayton C et al. The subcellular localisation of trypanosome RRP6 and its association with the exosome. Mol Biochem Parasitol 2007; 151:52-58.
24. Cristodero M, Böttcher B, Diepholz M et al. The exosome of Leishmania tarentolae: purification and structural analysis by electron microscopy. Mol Biochem Parasitol 2008; 159:24-29.
25. LaCava J, Houseley J, Saveanu C et al. RNA degradation by the exosome Is promoted by a nuclear polyadenylation complex. Cell 2005; 121:713-724.
26. Houseley J, Tollervey D. Yeast Trf5p is a nuclear poly(A) polymerase. EMBO Rep 2006; 7:205-211.
27. Cristodero M, Clayton C. Trypanosome MTR4 in involved in ribosomal RNA processing. Nucleic Acids Res 2007; 35:7023-7030.
28. Decuypere S, Vandesompele J, Yardley V et al. Differential polyadenylation of ribosomal RNA during posttranscriptional processing in Leishmania. Parasitology 2005; 131:321-329.
29. Kabani S, Fenn K, Ross A et al. Genome-wide expression profiling of in vivo-derived bloodstream parasite stages and dynamic analysis of mRNA alterations during synchronous differentiation in Trypanosoma brucei. BMC Genomics 2009; 10:427.
30. Queiroz R, Benz C, Fellenberg K et al. Transcriptome analysis of differentiating trypanosomes reveals the existence of multiple posttranscriptional regulons. BMC Genomics 2009; 10:495.
31. Jensen B, Sivam D, Kifer C et al. Widespread variation in transcript abundance within and across developmental stages of Trypanosoma brucei. BMC Genomics 2009; 10:482.
32. Haanstra J, Stewart M, Luu V-D et al. Control and regulation of gene expression: quantitative analysis of the expression of phosphoglycerate kinase in bloodstream form Trypanosoma brucei. J Biol Chem 2008; 283:2495-2507.
33. Mayho M, Fenn K, Craddy P et al. Post-transcriptional control of nuclear-encoded cytochrome oxidase subunits in Trypanosoma brucei: evidence for genome-wide conservation of life-cycle stage-specific regulatory elements. Nucleic Acids Res 2006; 34:5312-5324.
34. Gunzl A, Bruderer T, Laufer G et al. RNA polymerase I transcribes procyclin genes and variant surface glycoprotein gene expression sites in Trypanosoma brucei. Eukaryot Cell 2003; 2:542-551.
35. Hotz H-R, Biebinger S, Flaspohler J et al. PARP gene expression: regulation at many levels. Mol Biochem Parasitol 1998; 91:131-143.
36. Stoecklin G, Mayo T, Anderson P. ARE-mRNA degradation requires the 5′–3′ decay pathway. EMBO Rep 2006; 7:72-77.
37. Orban TI, Izaurralde E. Decay of mRNAs targeted by RISC requires XRN1, the Ski complex and the exosome. RNA 2005; 11:459-469.
38. Haile S, Dupe A, Papadopoulou B. Deadenylation-independent stage-specific mRNA degradation in Leishmania. Nucleic Acids Res 2008; 36:1634-1644.

39. Schwede A, Manful T, Jha B et al. The role of deadenylation in the degradation of unstable mRNAs in trypanosomes. Nucleic Acids Res 2009; 37:5511-5528.
40. Haile S, Estévez AM, Clayton C. A role for the exosome in the initiation of degradation of unstable mRNAs. RNA 2003; 9:1491-1501.
41. Schwede A, Ellis L, Luther J et al. A role for Caf1 in mRNA deadenylation and decay in trypanosomes and human cells. Nucleic Acids Res 2008; 36:3374-3388.
42. Li C-H, Irmer H, Gudjonsdottir-Planck D et al. Roles of a Trypanosoma brucei 5'->3' exoribonuclease homologue in mRNA degradation. RNA 2006; 12:2171-2186.
43. Irmer H, Clayton CE. Degradation of the EP1 mRNA in Trypanosoma brucei is initiated by destruction of the 3'-untranslated region. Nucleic Acids Res 2001; 29:4707-4715.
44. Anderson P, Kedersha N. RNA granules: posttranscriptional and epigenetic modulators of gene expression. Nat Rev Mol Cell Biol 2009; 10:430-436.
45. Buchan J, Parker R. Eukaryotic stress granules: the ins and outs of translation. Mol Cell 2009; 36:932-941.
46. Kramer S, Queiroz R, Ellis L et al. Stress granules and the heat shock response in Trypanosoma brucei. J Cell Sci 2008; 121:3002-3014.
47. Holetz F, Correa A, Avila A et al. Evidence of P-body-like structures in Trypanosoma cruzi. Biochem Biophys Res Commun 2007; 356:1062-1067.
48. Chowdhury A, Mukhopadhyay J, Tharun S. The decapping activator Lsm1p-7p–Pat1p complex has the intrinsic ability to distinguish between oligoadenylated and polyadenylated RNAs. RNA 2007.
49. Liu Q, Liang X, Uliel S et al. Identification and Functional Characterization of Lsm Proteins in Trypanosoma brucei. J Biol Chem 2004; 279:18210-18219.
50. Rodrıguez-Ezpeleta N, Brinkmann H, Burger G et al. Toward resolving the eukaryotic tree: the phylogenetic positions of Jakobids and Cercozoans. Curr Biol 2007; 17:1420-1425.

CHAPTER 5

THE EXOSOME AND 3′-5′ RNA DEGRADATION IN PLANTS

Heike Lange* and Dominique Gagliardi

Institut de Biologie Moléculaire des Plantes, UPR2357 du CNRS, Université de Strasbourg, Strasbourg, France.
**Corresponding Author: Heike Lange—Email: heike.lange@ibmp-cnrs.unistra.fr*

Abstract: One of the most versatile RNA degradation machines in eukaryotes is the 3'-5' RNA exosome. It consists of nine conserved subunits forming the core complex, which associates with active ribonucleases, RNA binding proteins, helicases and additional co-factors. While yeast and human exosome core complexes are catalytically inactive, the plant core complex has probably retained a phosphorolytic activity. Intriguingly, the down-regulation of individual subunits of the plant core complex in Arabidopsis mutants led to distinct developmental defects, suggesting an unequal contribution of the core subunits to the in vivo activities of the plant exosome complex. In addition, some of the plant core subunits as well as some associated factors are encoded by duplicated genes, which may have both overlapping and specific functions. Together, these results suggest an unique and complex organisation of exosome-mediated RNA degradation processes in plants. This chapter reviews our current knowledge of plant exosomes and discusses the impact of 3'-5' RNA degradation on the posttranscriptional control of plant genome expression.

INTRODUCTION

Exoribonucleolytic 3′-5′ RNA degradation makes a major contribution to RNA maturation, turnover and surveillance in all genetic compartments of plant cells. In chloroplasts and mitochondria, the major players are two organelle-specific polynucleotide phosphorylases (PNPases). In both the nucleus and the cytoplasm, the main 3′-5′ degradation machine is the exosome complex. RNA degradation by both PNPases and the exosome can be stimulated by the addition of oligo-A tails, a feature extensively discussed in Chapters 1, 6 and 8. PNPases and exosome complexes are structurally and evolutionary related. The present chapter will focus on some intriguing peculiarities that distinguish

RNA Exosome, edited by Torben Heick Jensen.

plant exosomes from their animal and fungal counterparts. At the end of the chapter we will discuss the impact of 3′-5′ degradation mechanisms to the posttranscriptional control of genome expression in the different compartments of plant cells

COMPOSITION OF PLANT EXOSOME CORE COMPLEXES

The eukaryotic exosome core complexes consist of three heterodimers of the PH-domain proteins RRP41-RRP45, RRP42-MTR3 and RRP43-RRP46 that form a ring-like structure, to which a "cap" of three S1/KH domain proteins, RRP4, RRP40 and CSL4, is bound (see Chapter 2 by K. Januszyk and C. Lima). Homologs of all nine core proteins are encoded in plants. Depending on the specie, a few proteins are encoded by duplicated genes. For instance, a short and a longer version of RRP45 (RRP45A and RRP45B) are encoded in the genomes of *Arabidopsis thaliana*, *Poplar trichocarpus* and *Glycine max*.[1] In contrast, *Vitis vinifera*, *Ricinus communis*, *Oryza sativa* and Zea mays seem to have either only the short or the long RRP45 protein. In addition, *Arabidopsis thaliana* harbours two nearly identical copies of RRP40 (termed AtRRP40A and AtRRP40B).[2]

The plant exosome complex was recently purified and characterised using transgenic *Arabidopsis* lines expressing tagged versions of AtRRP4 or AtRRP41, respectively. This revealed that the plant core exosome complex contains homologs of RRP4, CSL4, RRP41, RRP42, RRP43, RRP46 and MTR3. Under the experimental conditions used, only AtRRP40A and AtRRP45B were incorporated into the core complex.[2] Whether alternative core complexes containing AtRRP40B and/or AtRRP45A can be assembled, for instance in other tissues or other developmental stages, has not been determined yet. However, *Arabidopsis* single mutants lacking either AtRRP45 A or B have no phenotype or only a mild one, respectively, while simultaneous downregulation of both proteins is lethal.[1] Thus, it appears that AtRRP45A can partially complement for the loss of AtRRP45B. Moreover, expression of either AtRRP45A or AtRRP45B could restore the growth of a yeast *rrp45* null mutant. Together, these results suggest that AtRRP45A can indeed be incorporated into a functional core complex, at least under some circumstances.

Although composition and structural organisation of plant exosomes resemble those in other eukaryotes, plant exosomes have major peculiarities concerning the contribution of different core subunits to the in vivo activity of the exosome complex. In *Saccharomyces cerevisiae*, all nine subunits of the conserved core complex are essential.[3-5] Downregulation of individual subunits results in similar rRNA processing phenotypes, notably the characteristic accumulation of a 3′-extended transcript of the 5.8S rRNA. Moreover, the X-ray crystallographic analysis of the human exosome revealed unique interactions between all subunits of the core complex with each other.[6] It is therefore believed that all nine subunits are required to assemble a stable and functional core complex.[7] This appears to be different in *Arabidopsis*, because downregulation of the AtCSL4 subunit did not result in any obvious phenotype (Fig. 1) and size-fractionation from cells lacking AtCSL4 yielded nearly intact exosome complexes.[2] Interestingly, knock-down of the TbCSL4 subunit of *Trypanosome brucei* exosomes also did not result in degradation of other subunits, nor in disassembly of the core complex. Therefore, in both plants and trypanosomes, the CSL4 subunit of the cap appears to be dispensable for exosome function.[8,9] Interestingly, recent data indicate that the major part of CSL4 is also dispensable in yeast, because severely truncated versions of yeast CSL4, containing

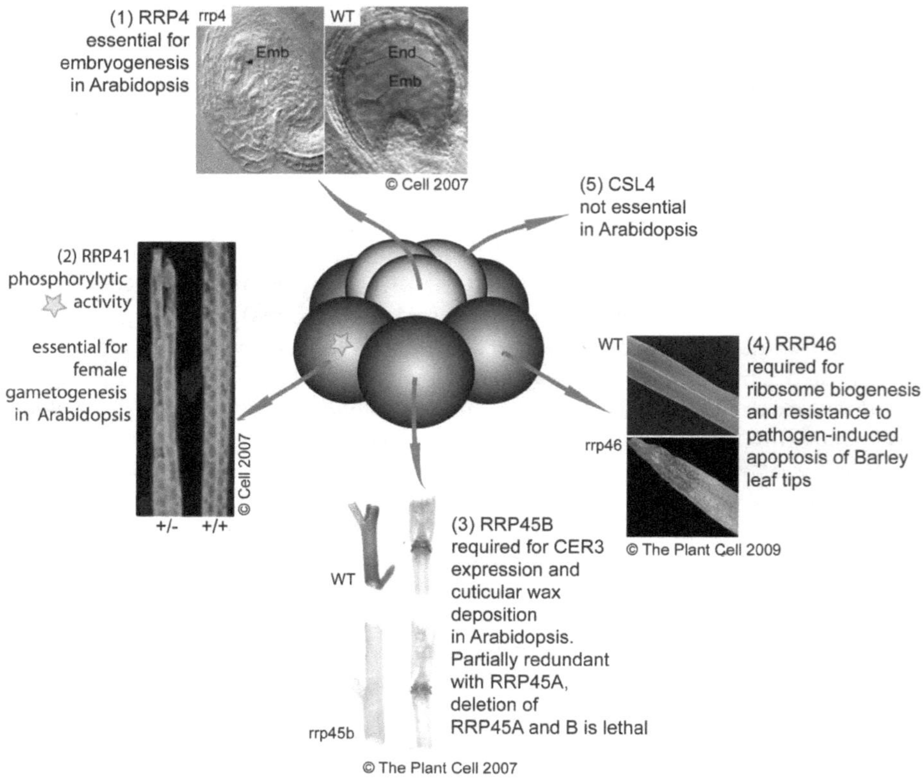

Figure 1. Phenotypes of plant RNA exosome mutants. The plant core exosome consists of six PH-domain proteins forming a ring-like structure (represented by dark grey balls) and three S1-domain proteins forming the cap (represented by light grey balls). Downregulation of distinct subunits of the core complex results in different defects of plant development: 1) Homozygous *rrp4* seeds arrest at a early stage of embryo development. Reprinted with permission from: Chekanova JA et al. Cell 2007; 131(7):1340-1353;[2] ©2007 Elsevier. 2) In contrast to other eukaryotic RNA exosomes where all subunits are catalytic inactive, the AtRRP41 subunit of the plant exosome has a phosphorolytic activity (indicated by the star). Heterozygous *rrp41/RRP41* plants produce viable seeds and aborted ovules (visible as empty spaces in siliques) in an 1:1 ratio. Reprinted with permission from: Chekanova JA et al. Cell 2007; 131(7):1340-1353;[2] ©2007 Elsevier. 3) RRP45B is required for proper expression of CER3, a protein of unknown function implicated in wax biosynthesis. Plants lacking RRP45B fail to express a CER3-GUS reporter gene (producing a blue indicator product) in stems and have defects in cuticular wax deposition. Reprinted with permission from: Hooker TS et al. Plant Cell 2007; 19(3):904-913;[1] ©2007 American Society of Plant Biologist. 4) Barley mutants lacking HvRRP46 are more sensitive to pathogen-induced apoptosis of leaf tips. Reprinted with permission from: Xi L et al. Plant Cell 2009; 21(10):3280-3295;[15] ©2009 American Society of Plant Biologists. 5) Arabidopsis plants lacking the CSL4 subunit do not display any obvious phenotype. These data indicate that subunits of the core complex have specialised roles in plant growth and development and make unequal contributions to the in vivo activity of the plant exosome.

either only the N-terminal RRP27 domain or the C-terminal Zn-ribbon-like domain, are sufficient to promote essential exosome functions.[10]

Another interesting aspect of plant exosomes concerns the enzymatic activity of the core complex. Structural-assisted sequence analysis and biochemical assays with purified or reconstituted exosome complexes indicated that in yeast, flies and animals, all nine subunits of the core complex are catalytically inactive.[6,11,12] By contrast, the critical residues of the catalytic site of the archaeal phosphorolytic RRP41 protein are conserved in AtRRP41. Moreover, recombinant AtRRP41 protein has a phosphorolytic activity in vitro.[13] If this activity is maintained upon complex assembly, plant exosomes would have a catalytic activity similar to archaeal exosomes (see Chapter 3 by E. Evguenieva-Hackenberg). Interestingly, the in vitro activity of recombinant AtRRP41 was stimulated after addition of short poly-A tails to the RNA substrates.[13] It is therefore possible that the exosome core complex itself degrades polyadenylated substrates in plant nuclei—unlike in yeast, where this activity depends on the nuclear exosome-associated exoribonuclease RRP6 (see next section).

FUNCTIONAL SPECIALISATION OF INDIVIDUAL SUBUNITS?

In contrast to the situation in yeast and trypanosomes, where depletion of individual subunits results in similar molecular phenotypes, depletion of individual subunits of the plant exosome give rise to distinct phenotypes on both the molecular level and on plant development.[2]

The first plant exosome mutant characterised was the *Arabidopsis cer7* mutant originating from a screen aimed at identifying wax biosynthesis genes (*eceriferum* lines).[14] *CER7* was found to be allelic to *RRP45B*,[1] encoding one of the PH-domain proteins copurifying with the exosome core complex.[2] Both plants carrying the *cer7* point mutation and plants harbouring a T-DNA insertion in the *RRP45B* gene have reduced levels of *CER3/WAX2* transcripts, which encode a protein of unknown function required to produce wild type levels of cuticular wax in *Arabidopsis* stems (Fig. 1).[1] This suggests that *CER7/AtRRP45B* acts as a positive regulator of *CER3* expression, probably by degrading transcripts of a putative *CER3* repressor. It is yet unclear if this function of AtRRP45B is linked to its function as an exosome subunit, or due to an exosome-independent activity of the protein. The RNA substrate (of either RRP45B or the exosome complex) that would explain the surprising specific phenotype of *cer7/rrp45b* mutants has not yet been identified. However, loss of RRP45B finally leads to a reduced cuticular wax load and reduced seed viability, but not to major defects in growth or development (Fig. 1). Plants lacking the closely related protein RRP45A have normal levels of cuticular wax, normal seed viability and do not exhibit any obvious phenotype, while the simultaneous disruption of *RRP45A* and *B* is lethal.[1] This indicates that the two RRP45 proteins have both distinct and overlapping functions. Furthermore, this result strongly suggests that at least one RRP45-like subunit is required for plant exosome assembly and/or function and, in conclusion, that functional exosomes are essential for plant viability. However, this view could be challenged by recent results obtained in *Hordum vulgare* (barley). While studying response reactions to the pathogen *Blumeria graminis* (abbreviated *bgh,* Barley-powdery Mildew), Xi et al isolated a barley mutant carrying a deletion of six genes, among them the gene encoding the exosome core subunit RRP46.[15] The *bcd1* (for *bgh*-induced tip cell death 1) mutation and virus-induced gene silencing of *RRP46* alone, resulted in rapid cell death in the top 2-3 cm of developing leaves after inoculation with the bacterial pathogen (Fig. 1). Pleiotropic symptoms such as a certain amount of sterile florets and a reduced number of tillers were also observed in both infected and non-infected plants. On the molecular level, loss of HvRRP46 leads to accumulation of polyadenylated, misprocessed rRNA precursors and, probably in a compensatory reaction, to a constitutive upregulation of ribosomal proteins. It was therefore suggested that the misregulation of ribosome biosynthesis pathways observed upon loss of HvRRP46 renders the plants more vulnerable to stress-induced cell-death.[15] However, these data suggest that HvRRP46, although clearly required for "classical" exosome-associated processing and degradation reactions, is not essential in barley. By contrast, it is also possible that a second RRP46-like protein with overlapping functions is encoded in the barley genome, or that one of the other subunits can replace HvRRP46 in the exosome core complex.

Interestingly, loss of AtRRP41 or AtRRP4, each encoded by a single gene in Arabidopsis, results in growth arrest at different stages of plant development (Fig. 1).[2] Heterozygous *rrp41/RRP41* mutants produce viable seeds and aborted ovules in an 1:1 ratio and the resulting progeny segregates 1:1 for wild type and heterozygous plants, indicating that *AtRRP41* is required for female gametogenesis (Fig. 1). By contrast, loss of *AtRRP4* does not affect pollen or ovule development, but severely impairs postzygotic processes because *rrp4* mutant seeds arrest in an early stage of embryo development with most seeds containing two-cell embryos and noncellularized endosperm (Fig. 1). However, downregulation of both *AtRRP4* or *AtRRP41* in seedlings using an inducible RNAi strategy leads to growth arrest and plant death, demonstrating that each of the two core subunits is also essential for post-embryonic growth.

In an exhaustive genome-wide tiling analysis Belostotsky and colleagues compared the polyadenylated RNA substrates that accumulated upon downregulation of AtRRP4 or AtRRP41 with those accumulating in *csl4* mutants.[2] They found that loss of AtCSL4 affected only a minor fraction of the RNA substrates affected upon loss of AtRRP4 and/or AtRRP41, suggesting that exosome complexes without AtCSL4 are still mostly functional. This study demonstrated also that downregulation of AtRRP4 or AtRRP41 affects overlapping, but also distinct pools of RNA substrates.[2] How can we explain that two essential subunits of the core complex make unequal contributions to the in vivo activity of the plant exosome? Apparently AtRRP41, or a RRP4-less core exosome, is sufficient to promote ovule development and a few cell divisions post fertilisation, while a fully assembled complex containing all nine conserved subunits is required to proceed into embryo development. One possible hypothesis is that AtRRP41, which in contrast to all other core subunits is catalytically active, can perform some functions as a monomer, or when assembled into the ring structure of the core exosome complex only. Another possibility is that alternative 9-subunit complexes, where RRP4 might be replaced by CSL4 or RRP40, can assemble and promote female gametogenesis and fertilisation, but not embryo development. It was therefore suggested that the ring and the cap structure may be less intertwined in plant exosomes than in yeast or human exosome complexes.[16]

Taken together, the available data strongly suggest that individual subunits of the Arabidopsis exosome core complex make unequal contributions to complex integrity and function. This challenges our all-or-nothing concept of the exosome complex and opens the possibility that distinct complexes with specialised functions can be assembled in plants. Interestingly, distinct cellular distribution patterns for individual subunits have been observed in *Drosophila melanogaster*, suggesting that complexes with different composition of subunits may also exist in flies.[17]

EXOSOME COFACTORS AND AUXILIARY PROTEINS IN PLANTS

Depending on the subcellular compartment, the core complex of yeast and animal exosomes associates with distinct exoribonucleases and other factors that confer both activity and specificity.[18,19] In yeast and drosophila both cytosolic and nuclear exosome preparations contain RRP44,[3,20] which has both an endonucleolytic and a hydrolytic 3′-5′ exoribonucleolytic activity (see Chapter 6 by A. Chlebowski and collegues).[10,21-23] Because the core exosome complex is catalytically inactive,[6,11,12] RRP44 is probably also responsible for the in vivo activity of the exosome of human and trypanosomes, despite the fact that RRP44 was not yet detected in exosome preparations from these organisms.

All plant genomes sequenced so far encode proteins with clear sequence similarity to RRP44. It appears that both dicots and monocots have two RRP44-like proteins, one containing and one lacking the PINc domain that confers the endonuclease activity in yeast and humans. However, no homologue of RRP44 was copurified with the *Arabidopsis* exosome complex,[2] and none of the plant proteins have been functionally characterised to date. Hence, the question whether and in which conditions one or both RRP44-like proteins can associate with the plant exosome and how this may influence substrate specificity and activity is unsolved.

Another protein that is stably associated with the nuclear exosome in yeast and with both nuclear and cytosolic exosomes in humans is RRP6/Pm-Scl100,[4] another hydrolytic exoribonuclease belonging to the RNase D family. The genomes of *Arabidopsis*, poplar and rice encode three RRP6-like (RRP6L) proteins.[2,24] RRP6L1 and RRP6L2 are more closely related to their yeast and animal counterparts, while RRP6L3 proteins appear to form a plant-specific subgroup. Interestingly, GFP-fusion protein analysis of stable *Arabidopsis* transformants revealed that the three RRP6-like proteins are localized in distinct subcellular or subnuclear compartments: RRP6L1 is detected mainly in the nucleoplasm, RRP6L2 was enriched in nucleoli and RRP6L3 appears to be cytoplasmic.[24] It remains unknown whether any of the three RRP6-like proteins can bind to the core complex, because no RRP6-like protein was copurified with the *Arabidopsis* exosome.[2] However, the nucleolar isoform RRP6L2 was shown to be involved in the degradation of the 5′ external transcribed spacer,[24] a maturation by-product of rRNA processing which accumulated also upon downregulation of either RRP41 or RRP4.[2] This indicates that RRP6L2 and the exosome core complex have at least some common substrates. Interestingly, the 5′ ETS appears not to be a substrate for RRP6L1, suggesting that the two nuclear RRP6-like proteins have distinct functions.[24]

Depending on the species and on experimental conditions, other proteins such as RNA binding proteins and RNA helicases have been detected in exosome preparations from yeast, flies, trypanosomes and humans. These cofactors include proteins of the cytosolic SKI-complex and the nuclear proteins RRP47/C1D/Lrp1 and MPP6 (see Chapter 8 by J.S. Butler and P. Mitchell).[9,25-31] In addition, some activities of the exosome require other cofactors that do not necessarily bind directly and/or stably to the core exosome complex, for instance the nuclear TRAMP (for TRF4/5-Air1/2-MTR4 -Polyadenylation) complex which is responsible for the oligo-adenylation of nuclear exosome substrates in *S. cerevisiae* and *Schizosaccharomyces pombe*.[32-34] In plants, no exosome cofactors have been characterised to date although candidates for most factors are present in *Arabidopsis*.

For instance, *Arabidopsis* expresses a large number of DEVD-box helicases, four of which share convincing sequence similarity with MTR4 and SKI2, respectively.[35] One of these helicases, ISE2, is involved in posttranscriptional gene-silencing and required for development of functional plasmodesmata during embryo development.[35] A second candidate, HEN2, is involved in flower organ identity.[36] Whether ISE2 or HEN2 participate in degradation processes mediated by the cytoplasmic exosome, or associate with other proteins resembling a SKI-like or TRAMP-like complex remains to be investigated. The fact that downregulation of exosome subunits or RRP6L2 results in the accumulation of polyadenylated substrates demonstrates the involvement of a noncanonical poly(A) polymerase in exosome-mediated RNA decay in plants.[2,24] Several candidate poly(A) polymerases are encoded in Arabidopsis, two of which have significant sequence similarity with the TRF4/5 proteins of the yeast TRAMP complex. One of these proteins, MEE44, appears to be essential for development of the female gametophyte alike AtRRP41, but has not been further characterised yet.[37,38]

Taken together, the contribution of associated exoribonucleases and other factors to the in vivo functions of the plant exosome is largely unknown. It remains possible that plant exosomes depend less on association with hydrolytic exoribonucleases, because the core complex may have retained a phosphorolytic activity that could be sufficient for some exosome functions.[2,13] Vice versa, RRP44 or RRP6-like proteins could also function independently of any physical association with the core complex. Such exosome-independent activities have already been described for yeast and drosophila RRP6.[39,40] The plant homologs of RNA helicases and poly(A)polymerases that function in exosome-mediated RNA decay and eventual plant-specific exosome cofactors remain to be identified.

RNA SUBSTRATES OF THE PLANT EXOSOME COMPLEX

Arabidopsis was the first organism whose exosome substrates have been characterised on a genome wide level. Using high-resolution tiling arrays covering the entire *Arabidopsis* genome, Chekanova and colleagues determined the polyadenylated RNA species that accumulated in seedlings upon RNAi-induced downregulation of AtRRP41 or AtRRP4, respectively.[2] This study identified about 1100 substrates probably of the nuclear exosome, since addition of short poly A tails tags nuclear transcripts for degradation.[18] Other potential exosome substrates such as deadenylated cytosolic mRNAs[41-43] or nonpolyadenylated nuclear transcripts, remain to be characterised.

About 300 transcripts accumulated either only in the RRP4-depleted or only in the RRP41-depleted sample. This confirmed that individual subunits of the core complex affect specific substrates and have different contributions to the in vivo activity of the exosome, as was already suggested by the distinct phenotype of the mutants. However, about 500 transcripts were upregulated in both samples and likely represent the common substrates of the exosome core complex.

Among those, a prominent group of transcripts were polyadenylated RNAs corresponding to rRNA precursor transcripts and maturation by-products removed during rRNA processing such as the external and internal transcribed spacer regions. This result confirmed that, as in other eukaryotes, one prominent function of the plant exosome is the removal of incompletely processed or misprocessed rRNA precursors and rRNA maturation by-products. In addition, plants downregulated for either RRP41 or RRP4 expression accumulate intermediates and by-products of miRNA processing, without affecting the levels of mature miRNAs. Interestingly, no tRNAs were observed, except tRNA-Tyr, a tRNA that undergoes multiple base modification steps during its maturation. It was therefore suggested that the plant exosome may participate in quality control of highly modified tRNAs as it was observed in yeast.[33]

In addition, other stable structural RNAs were identified as exosome substrates. These include some small nuclear RNAs (snRNA), most small nucleolar RNAs (snoRNA) and the RNA-polymerase III transcripts MRP/7-2 RNA and 7SL RNA. In most cases, both correctly processed and 3′-extended precursor transcripts accumulated as polyadenylated transcripts, suggesting that polyadenylation-assisted RNA decay by the plant exosome mediates both turnover of stable structural RNAs at the end of their lifetime and removal of misprocessed species. Alternatively, 3′ extended transcripts could represent maturation intermediates, i.e., incompletely processed precursor RNAs, since the exosome itself may trim the 3′ ends of some snoRNAs or other stable structural RNAs. The downregulation of plant exosome subunits also resulted in the upregulation of a number of mRNA loci, some of which accumulated either only in the RRP4-depleted or in the RRP41-depleted sample. However, as cytosolic mRNAs are assumed to undergo de-adenylation prior to degradation by the cytosolic exosome, it remains unclear whether the upregulated poly A^+ mRNAs found in this study represent true substrates of the exosome, or whether these mRNAs were upregulated due to secondary effects caused by the depletion of functional exosomes. In addition, a significant fraction of these upregulated regions corresponded to sense and antisense transcripts derived from potential (intronless) pseudogenes, or appeared to be aberrant readtrough transcripts from protein-coding genes.

Taken together, the results demonstrate that one of the major roles of the plant exosome is the degradation of all types of misprocessed and potentially nonfunctional transcripts including left-overs of nuclear or nucleolar RNA processing events. Moreover, plants depleted for RRP41 or RRP4 also accumulated a large variety of novel transcripts with no protein-coding potential and no predicted function.[2] In contrast to most of the above mentioned RNA species, these novel transcripts have extremely low steady-state levels and therefore were not detected by previous studies using wild type plants. A considerable fraction of these new RNAs derived from centromeric and pericentromeric heterochromatic regions that give also rise to small RNAs. Another group of short-lived noncoding transcripts of 100-600 nt corresponded to the 5′ regions of mRNAs. Similar noncoding RNAs, termed CUTs (cryptic unstable transcripts) or PROMPTs (promoter upstream transcripts) have been observed in yeast and mammals, respectively.[34,44,45] Yeast CUTs have been proposed recently to be global regulators of gene expression, because their synthesis might regulate transcription of neighbouring genes.[46,47] In plants, it remains to be explored to what extent these novel transcripts, the "dark matter of the transcriptome", have regulatory functions, or rather represent transcriptional noise. Understanding the nature and the origin of these spurious RNAs may give us some mechanistic insights into upstream processes as for example transcription initiation. However, the accumulation of this "hidden layer" of transcripts in absence of a functional exosome demonstrates that one of the important roles of the plant exosome is to counterbalance widespread transcription from heterochromatic and intergenic regions of the nuclear genome.

IMPACT OF 3′-5′ RNA DECAY ON PLANT GENOME EXPRESSION

The total number of about 1100 loci that were found upregulated in plant exosome mutants is not as high as could be expected for the main 3′-5′ degradation machine of a eukaryotic cell. Very likely, other RNA degradation pathways, in particular 5′-3′ exoribonucleolytic decay by XRN2, XRN3 and XRN4 can partially compensate for loss of exosome activity.[48] Vice versa, the cytosolic exosome complex may compensate for loss of the cytosolic 5′-3′ exoribonuclease XRN4, because only a relative small number of mRNAs were found upregulated in *xrn4* mutants.[49] A similar overlap of substrate specificity can be assumed for the nuclear compartment, as loss of the nuclear proteins XRN2 and XRN3 is associated with the accumulation of RNA excised from miRNA precursors,[50] similar to those that accumulate in exosome mutants.[2] Hence, it is very likely that in both cytosol and nucleus, a large proportion of transcripts can be degraded from either their 5′- or their 3′ end. In addition, when default degradation pathways are blocked, aberrant transcripts can enter silencing pathways that lead to the formation of small RNAs.[50-52]

In many cases this would subsequently result in endonucleolytic cleavage of the template by RISC, thereby generating new 3′ and 5′ ends accessible for both the exosome and 5′-3′ exoribonucleases. Indeed, it was shown that aberrant RNAs no longer removed by XRN4 become substrates of RNA-dependent RNA polymerases which results in the generation of small RNAs.[51,53] XRN2 and XRN3 also act as silencing suppressors.[50] In *S. pombe*, loss of the polymerase TRF4, which is responsible for the polyadenylation of exosome substrates, results in the generation of small RNAs predominantly from rRNAs and tRNAs.[54] These “false” small RNAs can compete with repeat-associated small RNAs for incorporation into Ago1 and have the potential to impair transcriptional silencing of heterochromatic and centromeric regions.[54] The overlap of exosome substrates with small-RNA generating loci suggests a similar relationship between silencing and 3′-5′ decay in plants.[2] However, a role of the plant exosome as silencing suppressor has not been directly demonstrated yet.

In conclusion, the impact of the nuclear exosome on plant RNA turnover, RNA quality control and nuclear RNA surveillance as revealed by the genome-wide tiling of exosome mutants is probably only the tip of the iceberg and 5′-3′ exoribonucleolytic degradation or silencing pathways, respectively, may have a similar broad and important impact on the final transcriptome of a plant cell.

IMPACT OF 3′-5′ DEGRADATION IN ORGANELLES

Polyadenylation-assisted RNA degradation exists also in the two other genetic compartments of plant cells, in chloroplast and mitochondria.[37,55] The main 3′-5′ exoribonucleases in plant organelles are chloroplast and mitochondrial polynucleotide phosphorylase (cpPNPase and mtPNPase), respectively.[56,55] PNPases are homo-trimeric enzymes containing six PH-domains that form a doughnut shape analogous to the ring structure of the exosome core complex. Alike the nuclear exosome, cpPNPase and mtPNPase degrade both coding and noncoding RNAs following polyadenylation of substrates. However, the impact of the poly(A)-assisted 3′-5′ RNA decay on the final transcriptome is remarkably different in the three compartments.[37] Downregulation of cpPNPase has only little effect on chloroplast mRNA abundance, suggesting that this pathway contributes to, but does not play a major role in controlling mRNA steady state levels.[57,58] Furthermore, most polyadenylation sites map to endonucleolytic cleavage sites, suggesting that endonucleases are important components of chloroplast RNA turnover.[59,60] In addition, a 5′-3′ RNA degradation pathway possibly involving homologs of RNase E and/or RNase J appears to have a more important impact on chloroplast mRNA levels than the 3′-5′ pathway.[61-65] To which extent the different pathways contribute to the removal of rRNA or maturation by-products has not been yet determined. Similarly, it is not known whether spurious transcription of intergenic regions, counterbalanced by rapid RNA decay, occurs in chloroplasts. By contrast, mitochondrial PNPase (mtPNPase) is essential in *Arabidopsis*.[66,67] Like the exosome in the nucleus, mtPNPase degrades antisense RNAs and cryptic transcripts that are generated from intergenic regions due to a relaxed transcriptional control and all sorts of nonfunctional RNAs such as maturation by-products of rRNA and tRNA processing.[56,68,69] In addition, plants lacking mtPNPase accumulate 3′ extended versions of mRNAs and rRNAs, suggesting that mtPNPase also mediates 3′ processing or quality control, or both.

While northern blots suggested that the accumulation of exosome substrates in *rrp41* and *rrp4* mutants is only moderate,[2] plants lacking mtPNPase accumulate tremendous high levels of some mitochondrial substrates.[56] This suggests that compensatory pathways exist in the nucleus, but are virtually absent in plant mitochondria. As a consequence, polyadenylation-assisted 3′-5′ RNA degradation by the exoribonuclease mtPNPase appears to be the main RNA processing and degradation pathway in plant mitochondria, with an enormous impact on the final mitochondrial transcriptome.

CONCLUSION

Structural and biochemical data obtained in other organisms strongly suggested an all-or-nothing mechanism for the eukaryotic exosome: all subunits are required to assemble a functional core complex, which, although it has no catalytic activity itself, serves as an essential assembly platform for active ribonucleases, RNA binding proteins and RNA helicases.[70-72] Hence, the most striking result of the recent characterisation of several plant exosome mutants is probably the unexpected unequal contribution of core subunits to the in vivo activity of the plant exosome.[1,2,15] Interestingly, depletion of individual exosome subunits in yeast and drosophila also affected both overlapping and distinct transcripts,[73,74] suggesting that a certain extent of functional specialisation may be a common feature of eukaryotic exosome subunits. Another unsolved issue is the contribution of cofactors to both activity and substrate specificity of exosome-mediated RNA decay in plants. The initial characterisation of the three RRP6L proteins showed that such potential cofactors can differ in their intracellular distribution and their substrate specificity.[24] Similarly, several Arabidopsis isoforms exists for most of the known exosome cofactors and to determine their intracellular localisation, their possible assembly with the exosome core or into other functional entities and their substrates will certainly help to better understand the complexity of 3′-5′ RNA degradation in plants. Finally, we need to understand the respective impact of 5′-3′ RNA degradation and silencing pathways to unravel how they are intertwined and to fully appreciate the impact of RNA degradation on plant growth and development.

REFERENCES

1. Hooker TS, Lam P, Zheng H et al. A core subunit of the RNA-processing/degrading exosome specifically influences cuticular wax biosynthesis in Arabidopsis. Plant Cell 2007; 19(3):904-913.
2. Chekanova JA, Gregory BD, Reverdatto SV et al. Genome-wide high-resolution mapping of exosome substrates reveals hidden features in the Arabidopsis transcriptome. Cell 2007; 131(7):1340-1353.
3. Mitchell P, Petfalski E, Shevchenko A et al. The exosome: a conserved eukaryotic RNA processing complex containing multiple 3′—>5′ exoribonucleases. Cell 1997; 91(4):457-466.
4. Allmang C, Petfalski E, Podtelejnikov A et al. The yeast exosome and human PM-Scl are related complexes of 3′—> 5′ exonucleases. Genes Dev 1999; 13(16):2148-2158.
5. van Hoof A, Parker R. The Exosome: A Proteasome for RNA? Cell 1999; 99(4):347-350.
6. Liu Q, Greimann JC, Lima CD. Reconstitution, Activities and Structure of the Eukaryotic RNA Exosome. Cell 2006; 127(6):1223-1237.
7. Lykke-Andersen S, Brodersen DE, Jensen TH. Origins and activities of the eukaryotic exosome. J Cell Sci 2009; 122(Pt 10):1487-1494.
8. Estévez AM, Kempf T, Clayton C. The exosome of Trypanosoma brucei. EMBO J 2001; 20(14):3831-3839.
9. Estévez AM, Lehner B, Sanderson CM et al. The roles of intersubunit interactions in exosome stability. J Biol Chem 2003; 278(37):34943-34951.
10. Schaeffer D, Tsanova B, Barbas A et al. The exosome contains domains with specific endoribonuclease, exoribonuclease and cytoplasmic mRNA decay activities. Nat Struct Mol Biol 2009; 16(1):56-62.
11. Liu Q, Greimann JC, Lima CD. Reconstitution, Activities and Structure of the Eukaryotic RNA Exosome. Cell 2007; 131(1):188-189.
12. Dziembowski A, Lorentzen E, Conti E et al. A single subunit, Dis3, is essentially responsible for yeast exosome core activity. Nat Struct Mol Biol 2007; 14(1):15-22.
13. Chekanova JA, Shaw RJ, Wills MA et al. Poly(A) tail-dependent exonuclease AtRrp41p from Arabidopsis thaliana rescues 5.8 S rRNA processing and mRNA decay defects of the yeast ski6 mutant and is found in an exosome-sized complex in plant and yeast cells. J Biol Chem 2000; 275(42):33158-33166.
14. Koornneef M, Hanhart CJ, Thiel F. A Genetic and Phenotypic Description of Eceriferum (cer) Mutants in Arabidopsis thaliana. J Hered 1989; 80(2):118-122.
15. Xi L, Moscou MJ, Meng Y et al. Transcript-based cloning of RRP46, a regulator of rRNA processing and R gene-independent cell death in barley-powdery mildew interactions. Plant Cell 2009; 21(10):3280-3295.
16. Belostotsky D. Chapter 21 Transcriptome Targets of the Exosome Complex in Plants. In: RNA Turnover in Eukaryotes: Nucleases, Pathways and Analysis of mRNA Decay.Vol Volume 448. Academic Press; 2008:429-443. Available at: http://www.sciencedirect.com/science/article/B7CV2-4V6YK6M-X/2/6fb63929ba1147e1f00cca6d57080bc1 [Accessed 2010].
17. Graham AC, Kiss DL, Andrulis ED. Differential distribution of exosome subunits at the nuclear lamina and in cytoplasmic foci. Mol Biol Cell 2006; 17(3):1399-1409.
18. Houseley J, LaCava J, Tollervey D. RNA-quality control by the exosome. Nat Rev Mol Cell Biol 2006; 7(7):529-539.
19. Schmid M, Jensen TH. The exosome: a multipurpose RNA-decay machine. Trends Biochem Sci 2008; 33(10):501-510.
20. Andrulis ED, Werner J, Nazarian A et al. The RNA processing exosome is linked to elongating RNA polymerase II in Drosophila. Nature 2002; 420(6917):837-841.
21. Lebreton A, Tomecki R, Dziembowski A et al. Endonucleolytic RNA cleavage by a eukaryotic exosome. Nature 2008; 456(7224):993-996.
22. Schneider C, Leung E, Brown J et al. The N-terminal PIN domain of the exosome subunit Rrp44 harbors endonuclease activity and tethers Rrp44 to the yeast core exosome. Nucleic Acids Res 2009; 37(4):1127-1140.
23. Mamolen M, Andrulis ED. Characterization of the Drosophila melanogaster Dis3 ribonuclease. Biochem Biophys Res Commun 2009; 390(3):529-534.
24. Lange H, Holec S, Cognat V et al. Degradation of a polyadenylated rRNA maturation by-product involves one of the three RRP6-like proteins in Arabidopsis thaliana. Mol Cell Biol 2008; 28(9):3038-3044.
25. Mitchell P, Petfalski E, Houalla R et al. Rrp47p is an exosome-associated protein required for the 3′ processing of stable RNAs. Mol Cell Biol 2003; 23(19):6982-6992.
26. Schilders G, Dijk EV, Pruijn GJM. C1D and hMtr4p associate with the human exosome subunit PM/Scl-100 and are involved in prerRNA processing. Nucleic Acids Res 2007; 35(8):2564-2572.
27. Schilders G, Raijmakers R, Raats JMH et al. MPP6 is an exosome-associated RNA-binding protein involved in 5.8S rRNA maturation. Nucleic Acids Res 2005; 33(21):6795-6804.
28. Milligan L, Decourty L, Saveanu C et al. A yeast exosome cofactor, Mpp6, functions in RNA surveillance and in the degradation of noncoding RNA transcripts. Mol Cell Biol 2008; 28(17):5446-57.

29. Cristodero M, Clayton CE. Trypanosome MTR4 is involved in rRNA processing. Nucleic Acids Res 2007; 35(20):7023-7030.
30. Brown JT, Bai X, Johnson AW. The yeast antiviral proteins Ski2p, Ski3p and Ski8p exist as a complex in vivo. RNA 2000; 6(3):449-457.
31. Hoof AV, Staples RR, Baker RE et al. Function of the ski4p (Csl4p) and Ski7p proteins in 3′-to-5′ degradation of mRNA. Mol Cell Biol 2000; 20(21):8230-8243.
32. LaCava J, Houseley J, Saveanu C et al. RNA degradation by the exosome is promoted by a nuclear polyadenylation complex. Cell 2005; 121(5):713-724.
33. Vanacova S, Wolf J, Martin G et al. A new yeast poly(A) polymerase complex involved in RNA quality control. PLoS Biol 2005; 3(6):e189.
34. Wyers F, Rougemaille M, Badis G et al. Cryptic pol II transcripts are degraded by a nuclear quality control pathway involving a new poly(A) polymerase. Cell 2005; 121(5):725-737.
35. Kobayashi K, Otegui MS, Krishnakumar S et al. INCREASED SIZE EXCLUSION LIMIT 2 encodes a putative DEVH box RNA helicase involved in plasmodesmata function during Arabidopsis embryogenesis. Plant Cell 2007; 19(6):1885-1897.
36. Western TL, Cheng Y, Liu J et al. HUA ENHANCER2, a putative DExH-box RNA helicase, maintains homeotic B and C gene expression in Arabidopsis. Development 2002; 129(7):1569-1581.
37. Lange H, Sement FM, Canaday J et al. Polyadenylation-assisted RNA degradation processes in plants. Trends Plant Sci 2009; 14(9):497-504.
38. Pagnussat GC, Yu H, Ngo QA et al. Genetic and molecular identification of genes required for female gametophyte development and function in Arabidopsis. Development 2005; 132(3):603-14.
39. Callahan KP, Butler JS. Evidence for core exosome independent function of the nuclear exoribonuclease Rrp6p. Nucl Acids Res 2008:gkn743.
40. Graham AC, Kiss DL, Andrulis ED. Core Exosome-independent Roles for Rrp6 in Cell Cycle Progression. Mol Biol Cell 2009; 20(8):2242-2253.
41. Hoof AV, Frischmeyer PA, Dietz HC et al. Exosome-mediated recognition and degradation of mRNAs lacking a termination codon. Science 2002; 295(5563):2262-2264.
42. Ibrahim F, Rohr J, Jeong W et al. Untemplated oligoadenylation promotes degradation of RISC-cleaved transcripts. Science 2006; 314(5807):1893.
43. Orban TI, Izaurralde E. Decay of mRNAs targeted by RISC requires XRN1, the Ski complex and the exosome. RNA 2005; 11(4):459-469.
44. Belostotsky D. Exosome complex and pervasive transcription in eukaryotic genomes. Curr Opin Cell Biol 2009; 21(3):352-358.
45. Preker P, Nielsen J, Kammler S et al. RNA exosome depletion reveals transcription upstream of active human promoters. Science 2008; 322(5909):1851-1854.
46. Neil H, Malabat C, d'Aubenton-Carafa Y et al. Widespread bidirectional promoters are the major source of cryptic transcripts in yeast. Nature 2009; 457(7232):1038-1042.
47. Xu Z, Wei W, Gagneur J et al. Bidirectional promoters generate pervasive transcription in yeast. Nature 2009; 457(7232):1033-1037.
48. Kastenmayer JP, Johnson MA, Green PJ. Analysis of XRN orthologs by complementation of yeast mutants and localization of XRN-GFP fusion proteins. Meth Enzymol 2001; 342:269-282.
49. Souret FF, Kastenmayer JP, Green PJ. AtXRN4 degrades mRNA in Arabidopsis and its substrates include selected miRNA targets. Mol Cell 2004; 15(2):173-183.
50. Gy I, Gasciolli V, Lauressergues D et al. Arabidopsis FIERY1, XRN2 and XRN3 are endogenous RNA silencing suppressors. Plant Cell 2007; 19(11):3451-3461.
51. Gregory BD, O'Malley RC, Lister R et al. A link between RNA metabolism and silencing affecting Arabidopsis development. Dev Cell 2008; 14(6):854-866.
52. Voinnet O. Use, tolerance and avoidance of amplified RNA silencing by plants. Trends in Plant Science 2008; 13(7):317-328.
53. Gazzani S, Lawrenson T, Woodward C et al. A link between mRNA turnover and RNA interference in Arabidopsis. Science 2004; 306(5698):1046-1048.
54. Bühler M, Spies N, Bartel DP et al. TRAMP-mediated RNA surveillance prevents spurious entry of RNAs into the Schizosaccharomyces pombe siRNA pathway. Nat Struct Mol Biol 2008; 15(10):1015-1023.
55. Schuster G, Stern D. RNA polyadenylation and decay in mitochondria and chloroplasts. Prog Mol Biol Transl Sci 2009; 85:393-422.
56. Holec S, Lange H, Kühn K et al. Relaxed transcription in Arabidopsis mitochondria is counterbalanced by RNA stability control mediated by polyadenylation and polynucleotide phosphorylase. Mol Cell Biol 2006; 26(7):2869-2876.
57. Walter M, Kilian J, Kudla J. PNPase activity determines the efficiency of mRNA 3′-end processing, the degradation of tRNA and the extent of polyadenylation in chloroplasts. EMBO J 2002; 21(24):6905-6914.

58. Komine Y, Kikis E, Schuster G et al. Evidence for in vivo modulation of chloroplast RNA stability by 3′-UTR homopolymeric tails in Chlamydomonas reinhardtii. Proc Natl Acad Sci USA 2002; 99(6):4085-4090.
59. Lisitsky I, Klaff P, Schuster G. Addition of destabilizing poly (A)-rich sequences to endonuclease cleavage sites during the degradation of chloroplast mRNA. Proc Natl Acad Sci USA 1996; 93(23):13398-13403.
60. Kudla J, Hayes R, Gruissem W. Polyadenylation accelerates degradation of chloroplast mRNA. EMBO J 1996; 15(24):7137-7146.
61. Mudd EA, Sullivan S, Gisby MF et al. A 125 kDa RNase E/G-like protein is present in plastids and is essential for chloroplast development and autotrophic growth in Arabidopsis. J Exp Bot 2008; 59(10):2597-2610.
62. Drager RG, Higgs DC, Kindle KL et al. 5′ to 3′ exoribonucleolytic activity is a normal component of chloroplast mRNA decay pathways. Plant J 1999; 19(5):521-531.
63. Loiselay C, Gumpel NJ, Girard-Bascou J et al. Molecular identification and function of cis- and trans-acting determinants for petA transcript stability in Chlamydomonas reinhardtii chloroplasts. Mol Cell Biol 2008; 28(17):5529-5542.
64. Pfalz J, Bayraktar OA, Prikryl J et al. Site-specific binding of a PPR protein defines and stabilizes 5′ and 3′ mRNA termini in chloroplasts. EMBO J. 2009. Available at: http://www.ncbi.nlm.nih.gov/pubmed/19424177 [Accessed 2009].
65. Johnson X, Wostrikoff K, Finazzi G et al. MRL1, a Conserved Pentatricopeptide Repeat Protein, Is Required for Stabilization of rbcL mRNA in Chlamydomonas and Arabidopsis. Plant Cell 2010. Available at: http://www.ncbi.nlm.nih.gov/pubmed/20097872 [Accessed 2010].
66. Perrin R, Lange H, Grienenberger J et al. AtmtPNPase is required for multiple aspects of the 18S rRNA metabolism in Arabidopsis thaliana mitochondria. Nucleic Acids Res 2004; 32(17):5174-5182.
67. Perrin R, Meyer EH, Zaepfel M et al. Two exoribonucleases act sequentially to process mature 3′-ends of atp9 mRNAs in Arabidopsis mitochondria. J Biol Chem 2004; 279(24):25440-25446.
68. Holec S, Lange H, Canaday J et al. Coping with cryptic and defective transcripts in plant mitochondria. Biochim Biophys Acta 2008; 1779(9):566-573.
69. Holec S, Lange H, Dietrich A et al. Polyadenylation-mediated RNA degradation in plant mitochondria. Methods Enzymol 2008; 447:439-461.
70. Vanacova S, Stefl R. The exosome and RNA quality control in the nucleus. EMBO Rep 2007; 8(7):651-657.
71. Lebreton A, Séraphin B. Exosome-mediated quality control: substrate recruitment and molecular activity. Biochim Biophys Acta 2008; 1779(9):558-565.
72. Houseley J, Tollervey D. The many pathways of RNA degradation. Cell 2009; 136(4):763-776.
73. Houalla R, Devaux F, Fatica A et al. Microarray detection of novel nuclear RNA substrates for the exosome. Yeast 2006; 23(6):439-454.
74. Kiss DL, Andrulis ED. Genome-wide analysis reveals distinct substrate specificities of Rrp6, Dis3 and core exosome subunits. RNA 2010. Available at: http://www.ncbi.nlm.nih.gov/pubmed/20185544 [Accessed 2010].

CHAPTER 6

CATALYTIC PROPERTIES OF THE EUKARYOTIC EXOSOME

Aleksander Chlebowski,[1] Rafał Tomecki,[1] María Eugenia Gas López,[2] Bertrand Séraphin[2] and Andrzej Dziembowski*,[1]

[1]Institute of Genetics and Biotechnology, Faculty of Biology, University of Warsaw, and Institute of Biochemistry and Biophysics, Polish Academy of Sciences, Warsaw, Poland; [2]Insitute of Genetics and Molecular and Cellular Biology, Illkirch, France.

**Corresponding Author: Andrzej Dziembowski—Email: andrzejd@ibb.waw.pl*

Abstract: The eukaryotic exosome complex is built around the backbone of a 9-subunit ring similar to phosporolytic ribonucleases such as RNase PH and polynucleotide phosphorylase (PNPase). Unlike those enzymes, the ring is devoid of any detectable catalytic activities, with the possible exception of the plant version of the complex. Instead, the essential RNA decay capability is supplied by associated hydrolytic ribonucleases belonging to the Dis3 and Rrp6 families. Dis3 proteins are endowed with two different activities: the long known processive 3′-5′ exonucleolytic one and the recently discovered endonucleolytic one. Rrp6 proteins are distributive exonucleases. This chapter will review the current knowledge about the catalytic properties of theses nucleases and their interplay within the exosome holocomplex.

INTRODUCTION

When the exosome was first described, it was as a “multienzyme ribonuclease complex” and three subunits of the yeast complex, Rrp4, Rrp41 and Dis3, were shown to be active ribonucleases in vitro.[1] The study demonstrated three different RNase activities within the complex: a distributive hydrolytic one for Rrp4, a processive phosphorolytic one for Rrp41 and a processive hydrolytic one for Dis3. Notably, the three modes of RNA degradation were ascribed to proteins that are vastly different: Rrp41 is similar to the bacterial phosphorolytic enzyme, RNase PH, Dis3 is a member of the RNR family of hydrolytic ribonucleases, whereas Rrp4 contains no domains

RNA Exosome, edited by Torben Heick Jensen.

with known RNase activity. The latter led to Rrp4 becoming the founding member of a new superfamily of ribonucleases.[2]

As more components of the exosome were discovered, they were assumed to be ribonucleases based on sequence similarity to Rrp4 and Rrp41.[3] Expectations grew for the complex to be a very complicated machine with numerous tools for the destruction of RNAs and speculations began about how the many activities are spatially organized, selected for action and coordinated. Structurally, the exact protein stoichiometry was still unknown, it was even unclear, whether the exosome population is homogenous or not, that is whether all exosome particles share the same subunit composition. And if not, whether the composition of particular complexes is constant or dynamic.

The first clues to the structure of the exosome came relatively early, from analyses of archaeal genomes.[4] It was revealed that most of them encode only two proteins homologous to RNase PH, while no less than six such proteins are encoded in eukaryotic genomes. It was then proposed that, as in the case of the proteasome, the same multimeric structure is assembled using either multiple copies of two proteins or individual copies of six different ones. This pointed to the exosome consisting of, among others, six proteins with the RNase PH fold and three containing the S1 domain or multiples thereof.

RNase PH and S1 domains with a 6:3 stoichiometry were also found in the trimeric structure of *Streptomyces antibioticus* PNPase[5] and thus the yeast exosome was predicted to assume a related architecture, which was confirmed by electron-microscopy studies.[6] These observations were followed by a steady flow of structural data on archaeal and yeast exosomes, strengthening the notion that the exosome is homogenously built around a structure of defined composition.[7-9] Finally the crystal structure of the human ring complex was solved, giving the first high-resolution view of a 9-subunit exosome complex.[10] Recently a crystal structure was also obtained for yeast Dis3 bound to some ring subunits.[11]

As the structural characterization of the exosome progressed, knowledge about its catalytic properties was revolutionized. Solving the structure of the human ring complex[10] coincided with the contributions of particular subunits to exosome activity being carefully reevaluated.[12] Both studies showed beyond doubt that the 9-subunit ring complex is catalytically inactive in yeast and human. The current view of the exosome is thus quite different from the initial one: only a few proteins in the complex are active RNases. The precise function of the remaining majority is elusive but their involvement in substrate recruitment is often suggested and a recent study indicates a role in substrate selection.[11]

EXOSOME COMPOSITION

The eukaryotic exosome is composed of a ring-like structure and accessory subunits.* The ring consists of nine proteins: six composed of a single domain homologous to

* The name "exosome core" is used to describe several structural elements. Most commonly, it refers to the 9-subunit complex, archaeal or eukaryotic, but has also been used to mean the hexameric ring of RNase PH subunits of Archaea, as well as the 10-subunit complex of eukaryotes, encompassing the 9-subunit ring and Dis3. *To avoid confusion we shall refrain from using "exosome core" throughout the chapter and instead use the term "exosome ring" to denote the 9-subunit complex of six RNase PH subunits and three S1/KH subunits.*

the phosphorolytic enzyme RNase PH and three containing RNA binding domains: S1 and KH. The RNase PH subunits form a toroidal hexamer with the RNA binding subunits residing on one side (considered to be the "top") and a channel running along the pseudosymmetry axis of the complex.[10] The RNase PH-like subunits are: Rrp41, Rrp42, Rrp43, Rrp45, Rrp46 and Mtr3 and the RNA binding subunits are: Rrp4, Rrp40 and Csl4.

Despite extensive similarity to RNase PH all of the ring subunits are catalytically inactive in yeast and human, having lost the key catalytic amino acid residues of the bacterial enzyme.[10,12] The enzymatic activities detected previously in Rrp4 and Rrp41 were apparently artefactual. Of all eukaryotic exosome ring proteins studied so far, only plant Rrp41 homologues seem capable of catalysis. Indeed, Rrp41 of *Arabidopsis thaliana* has been convincingly shown to have phosphorolytic activity,[13] indicating that exosome mediated RNA turnover pathways in plants may differ substantially from those of other organisms. This is covered in detail in Chapter 5 by H. Lange and D. Gagliardi.

The exosome ring associates with additional subunits that supply catalytic activity to the complex. In yeast these are the essential Dis3, present both in the nucleus and the cytoplasm and the non-essential Rrp6, which is found only in the nucleus.[14] Thus, the yeast exosome exists in two forms: the 10-subunit cytoplasmic and the 11-subunit nuclear form, both of which interact with more factors and protein complexes in their respective compartments. This model of exosome composition was thought to be ubiquitous among eukaryotes until some recent studies have challenged this assumption, as more catalytic exosome proteins were discovered.

The human genome encodes two proteins of the Dis3 family, named DIS3 and DIS3-like (DIS3L) and the same appears to be true for all vertebrates. DIS3, the closest homologue of yeast Dis3, is localized mostly, if not exclusively, in the nucleus, while DIS3L is strictly cytoplasmic. It would seem the nuclear and cytoplasmic functions of yeast Dis3 have been divided between two proteins in vertebrates. Notably, in human cells DIS3 is absent from nucleoli, unlike the yeast protein, which is enriched in this compartment.[15,16] Interestingly, the single human Rrp6 protein is found not only in the nucleus (with nucleolar enrichment typical of exosome ring proteins) but also in the cytoplasm, unlike its yeast counterpart.[15,16]

In plants on the other hand, while only one protein of the Dis3 family is encoded in currently known plant genomes, three genes of the *RRP6* family are present in *Arabidopsis*, rice and poplar, encoding the proteins AtRRP6L1, AtRRP6L2 and AtRRP6L3 in *Arabidopsis*. Like the vertebrate Dis3 paralogues, the plant Rrp6 proteins are differentially localized in the cell: AtRRP6L1 and AtRRP6L2 were both found to localize in the nucleus, although with different distributions, whereas AtRRP6L3 was exclusively cytoplasmic.[17] The subcellular localization of Dis3 in plants has not been determined. Neither have physical interactions between the plant exosome ring and any of the putative catalytic subunits been observed so far.[18] Thus the interplay between the catalytic subunits of the exosome may be quite different between eukaryotic taxa. The different isoforms of the exosome are depicted in Figure 1.

Notably, genomes of eukaryotes, including plant and vertebrate, do seem to encode other proteins of the RNR family than Dis3, but these are unlikely to physically interact with the exosome, given their lack of the PIN domain (see below). Their potential role in RNA metabolism notwithstanding, they are not considered to be exosome proteins.

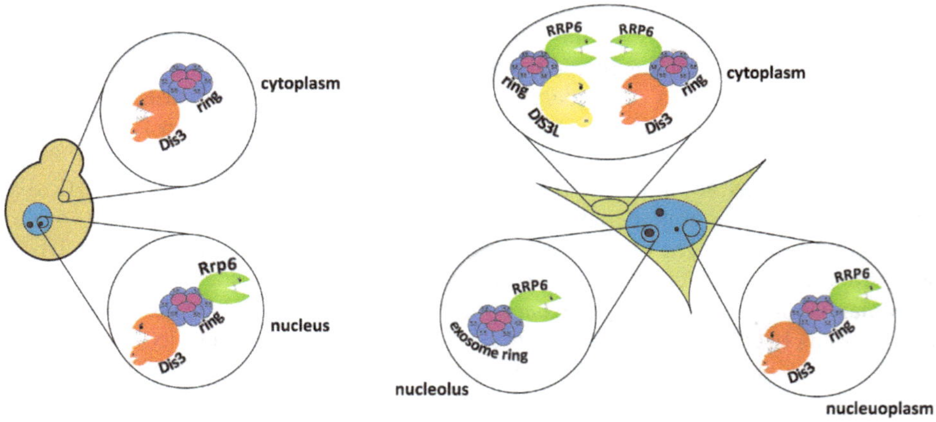

Figure 1. Composition of yeast and human exosome complexes and differential localization of catalytic subunits. The exosome ring is localized in the cytoplasm, nucleus and nucleoli in all eukaryotes. In *Saccharomyces cerevisiae* (left) the single Dis3 protein is present in all three compartments and the single Rrp6 protein is confined to the nucleus. In *Homo sapiens* (right) the single RRP6 homologue is found in all compartments. One Dis3 paralogue—DIS3—is mostly, if not only, nuclear, but excluded from nucleoli, while the other—DIS3L—is strictly cytoplasmic.

DIS3

Dis3 proteins are highly conserved in all eukaryotes and although there are differences at the sequence level, in terms of domain organization they are all identical. The generic Dis3 protein is a representative of the RNR superfamily of ribonucleases, with similarity to bacterial RNase II and RNase R. Like the bacterial enzymes, it consists of three RNA binding domains (two cold-shock (CSD) and one S1) and one catalytic domain (RNB). However, unlike the bacterial RNases, Dis3 additionally has a long N-terminal region containing a PIN domain that is responsible for stable association of the protein with the exosome ring. This is absolutely crucial for Dis3 to function as part of the exosome complex.[11,19] The domain composition of Dis3 is shown in Figure 2.

Enzymes of the RNR family all exhibit hydrolytic activity against RNA, which they digest processively in the 3′-5′ direction. When presented with a substrate, they release nucleoside 5′-monophosphates and end products: short oligonucleotides, 2-5 nt in length, depending on the particular enzyme.[12,20] However, they can have quite different substrate specificities. For instance, RNase II acts only on single-stranded RNA and stops a few nucleotides before structured regions, RNase R on the other hand is able to digest through both intra- or intermolecular RNA secondary structures, provided a single-stranded region at the 3′ end is available for the protein to bind.[20] All RNR proteins require the presence of divalent metal ions for catalysis, preferably magnesium.

RNB is a well characterized domain. Crystal structures of substrate-bound *Escherichia coli* RNase II and *Saccharomyces cerevisiae* Dis3 reveal the structural basis for magnesium ion dependent hydrolytic RNA cleavage as well as for the substrate specificities of the enzymes. The active site lies buried deep within the domain and is connected to the surface by a narrow channel that can only accommodate single-stranded substrates. Two magnesium ions are coordinated in the active site by conserved aspartic acid residues and mutation of one of those residues results in complete loss of catalytic activity.[21,22]

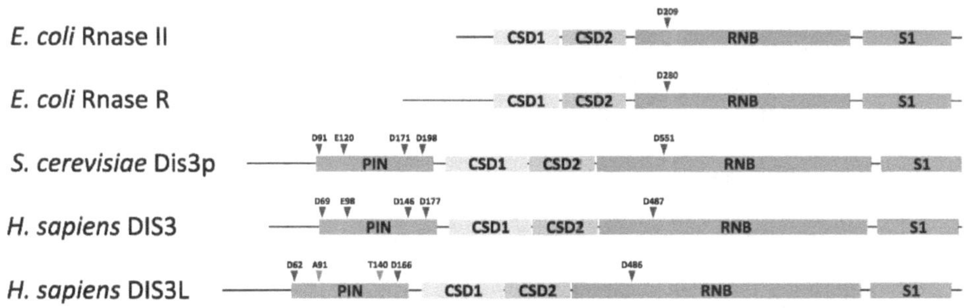

Figure 2. Domain organization of the RNR superfamily. All RNR proteins contain one catalytic domain, RNB and three RNA binding domains: CSD1, CSD2 and S1. Proteins of the Dis3 subfamily, found in all eukaryotes but never in prokaryotes, also contain a catalytic PIN domain. The latter is inactive in the Dis3l proteins, missing two of the conserved acidic amino acid residues (light gray/blue arrows).

The PIN domain is found in proteins of different functions and was initially associated with signaling pathways[23] but some PIN domains were later predicted and shown to have nuclease activity.[24,25] Cleavage of RNA by the PIN domains in vitro requires the presence of divalent metal ions, usually manganese.[26]

Crystal structures of different PIN domains show an RNase H-like fold with the catalytic site placed in a ridge running along the surface of the domain.[24] Metal ions are coordinated in the active site by four conserved acidic amino acid residues and mutation of any of these residues abolishes RNase activity.[24,26] Interestingly, some PIN domains act as oligomers with different catalytic properties.[24]

Exonuclease Activity

Since the first demonstration that yeast Dis3 is an exoribonuclease[1] the exonucleolytic activity of Dis3 has been exhaustively assayed and while the majority of studies concerned the activity of the yeast protein, those of the *Drosophila melanogaster* and human homologues have also been described.[12,15,16,27] As expected of an RNR protein, Dis3 is a processive 3′-5′ exoribonuclease that requires divalent metal ions, releases 5′-mononucleotides and leaves a short end product.

Maximum activity of Dis3 in vitro is observed for unusually low concentrations of magnesium ions, about 50 μM.[12] This is a little puzzling, given that the intracellular concentrations of this cation are more than ten times higher.[28] Apparently the cell "intentionally" keeps exosome activity low by maintaining a suboptimal environment or perhaps the activity is enhanced by some unidentified cofactors. However, such results and conclusions should be taken with a pinch of salt. One must keep in mind that determining actual amounts of free magnesium ions in relevant environments is difficult: magnesium is bound by free trinucleotides and RNA and these species are very abundant in cells as well as in enzymatic assay reaction solutions. For these reasons we will concentrate more on qualitative results of activity studies than on quantitative ones.

Dis3 is active on various substrates, but does exhibit some specificity when degradation efficiency is considered. The strongest activity was observed for AU-rich substrates, which may reflect a role of the exosome in degrading ARE-containing mRNAs.[29,30] Activity on adenine homopolymers or oligoadenylated heteropolymers was more modest, but they were still degraded completely.[10,22]

Importantly, Dis3 is able to degrade structured substrates, making it biochemically more similar to RNase R than to RNase II. Like RNase R, Dis3 appears to bind RNA very efficiently and even though the rate of hydrolysis drops significantly on structured regions, the protein does not dissociate and degrades the molecule to end products.[10]

Degradation of Structured Substrates

Dis3 is one of the very rare exonucleases that can degrade both single- and double-stranded substrates. The enzyme cannot bind blunt-ended dsRNA molecules, but rather needs a single stranded 3′ end.[22] Once bound, Dis3 begins processive hydrolysis and even if secondary structures are encountered, the protein does not dissociate, slowly moving through the structured region. These RNA unwinding properties apply to both inter- and intramolecular secondary structures.[10,12,22]

Like RNase II, the RNR part of yeast Dis3 has also been crystallized and its structure solved, giving more insight into the *modus operandi* of the RNR family of nucleases. Both proteins were expressed as mutants unable of catalysis (RNase II^{D209N} and Dis3^{D551N}) and crystallized in the presence of ssRNA and both show that the active site of the RNB domain is situated at the bottom of a narrow channel deep enough to accomodate 5 or 6 nucleotides.[21,22] Comparison with RNase II reveals strong structural conservation within all individual domains, but a markedly different spatial arrangement of the RNA binding domains. This results in the substrate strand taking a different route on its way to the inward channel of the RNB domain: in RNase II the substrate is threaded between the CSD2 and S1 domains only (Fig. 3A), while in Dis3 it is threaded between the CSD1 and RNB domains, contacting different surfaces (Fig. 3B). It has been suggested that this different positioning of domains and the resulting bending of the RNA strand by Dis3 is responsible for the difference between RNase II and RNase R-like activity on structured RNA substrates. Specific point mutations, designed according to the structure model, impaired Dis3 activity on double-stranded, but not on single-stranded RNA, lending support to this hypothesis.[22] But that is not the end of it. The problem has also been tackled by a comparative study of RNase II and RNase R, in which the contribution of individual domains to enzymatic activity was measured. Curiously, the difference between the two bacterial enzymes lies within the RNB domain itself, whereas the CSD and S1 domains, while important for substrate binding and greatly enhancing the catalytic properties of the proteins, are not essential and have no bearing on substrate specificities.[31] The same authors demonstrated that high-affinity binding of substrate by the RNB domain of RNase R allows the enzyme to take advantage of thermal breathing of RNA duplexes and thus move along and digest structured substrates.[32]

The two models, while not strictly exclusive, seem contradictory. Either one of them is wrong or RNase R and Dis3 have developed completely different strategies to achieve the same result. Such functional convergence would be very interesting since the proteins are closely related, but it seems more likely that the underlying mechanism of structured substrate digestion is the same throughout the RNR superfamily. In an attempt to reconcile the two hypotheses it has been suggested that the positioning of the single RNA strand in the Dis3 crystal is in fact artefactual.[32]

According to this model, in an RNase R-like enzyme there are two paths for RNA to take: one leading between the CSD2 and S1 domains toward the channel entrance and the other leading away from the channel entrance, between the CSD1 and RNB domains and outward. The free 3′ end of an otherwise structured substrate would approach the

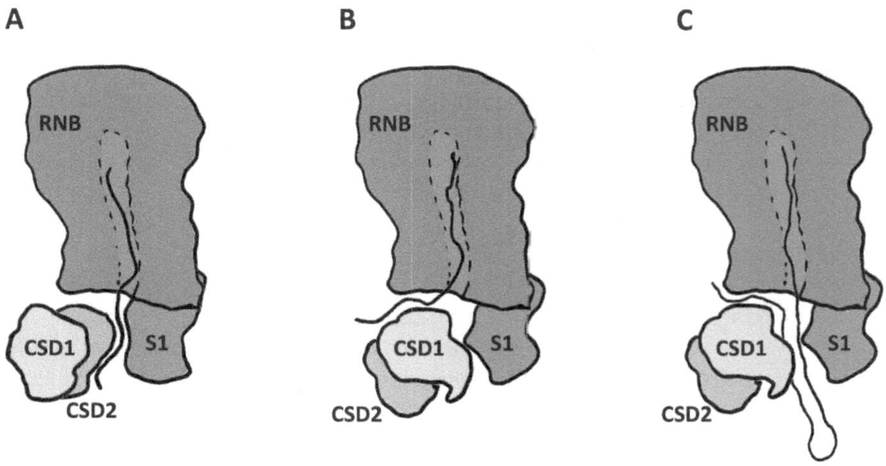

Figure 3. Unwinding secondary structures by Dis3. A) Substrate position observed in crystals of *Escherichia coli* RNase II. The RNA approaches the RNB channel between the RNA binding domains, making contacts with CSD2 and S1. B) Substrate position observed in crystals of *Saccharomyces cerevisiae* Dis3. The RNA takes a different route, between CSD1 and RNB. C) Proposed position of a structured RNA molecule in Dis3. The 3′ portion of the substrate reaches the catalytic site and is processively hydrolyzed. The double stranded segment approaches the RNB channel between the RNA binding domains, but is too wide to enter RNB itself. As the 3′ end is pulled into the active site with each successive cleavage event, the secondary structure is pulled apart, the 3′ nucleotide entering the RNB channel and the 5′ nucleotide leaving the protein between RNB and CSD1. Presumably, once the secondary structure is completely unwound, the 5′ portion of the RNA is pulled back and threaded into the RNB channel to be completely digested.

RNB domain by the first path, as observed for RNase II.[21] When the double-stranded region would reach the entrance to the channel, the enzyme would cleave off the last easily accessible nucleotide at the 3′-end and then pause but not dissociate, due to high affinity of the RNB domain towards RNA.[31] A wedge is then postulated to protrude from the surface of the RNB domain and push into the double-stranded region of the RNA, disrupting the structure and enhancing thermal breathing effects. Upon transient melting of the secondary structure, the enzyme would immediately move along the substrate, driven by the thermodynamically favourable filling of the active site by the 3′ terminus of the RNA and push the other strand into the second path. Thus the double-stranded region would be passively unwound, split on the edge of the channel in the RNB domain (Fig. 3C).

The hypothesis is supported by several observations. First, activity of RNase R against structured substrates depends on temperature to a greater extent than its activity against ssRNA, suggesting involvement of thermal breathing.[32] Second, the RNB domain of RNase R binds RNA much stronger than the RNB domain of RNase II, probably by making strong contacts between the RNA backbone and the walls of the channel.[31] Only two such contacts are seen in the crystal structure of RNase II.[21] Furthermore, an RNase R mutant, in which some residues within the channel are changed to those found in RNase II have catalytic properties similar to those of RNase II, including faster action on unstructured substrates and problems with structured ones.[32] Third, blocking the second path by introducing a bulky amino acid side chain makes it harder for Dis3 to digest structured substrates with no effect on its activity towards unstructured ones. Such blockage occurs naturally in RNase II.[22] Finally, the second path is blocked in the crystal structure of yeast Dis3 bound to ring components (Rrp41 and Rrp45), while the first remains open and a ring+Dis3 complex is still able to degrade structured substrates.[11] The latter not only is consistent with the thermal breathing-driven mechanism of RNA unwinding by pulling one strand into a narrow channel, but also indicates that secondary structures can be split on other surfaces than the RNB domain itself.

Endonuclease Activity

Recently three groups independently discovered an additional endoribonuclease activity of yeast Dis3, which resides in the PIN domain.[19,33,34] The protein was able to digest circular substrates as well as linear ones and the latter produced similar patterns of products regardless of which end of the substrate was labeled. No cleavage was observed in double-stranded regions of structured substrates.[19] The endoribonuclease activity was dependent on divalent metal ions, highest with manganese, but also supported by zinc and, to a small extent, magnesium.

The endonuclease activity of the PIN domain is unspecific to substrate sequence, but has been reported to be specific towards phosphorylated 5′ ends. When linear substrates were used, molecules with 5′ phosphate groups were cleaved very efficiently, whereas ones with 5′ hydroxyl groups were not.[34] However, another group was unable to reproduce this result and observed comparable activity for 5′-phosphorylated and -unphosphorylated substrates (Rafał Tomecki, unpublished data). Phosphorylation of the 3′ terminus is irrelevant for the endoribonuclease activity.[34]

Paralogy in the Dis3 Family

Vertebrate cells differ from others in that they contain two Dis3 proteins, called Dis3 and Dis3l, most likely resulting from a gene duplication event. The two subfamilies have diverged significantly: in terms of sequence similarity Dis3l proteins are equally distant from their respective Dis3 paralogue and from Dis3 of *Saccharomyces cerevisiae*. Nevertheless, the domain structure has remained the same in both paralogous lines. The human proteins, DIS3 and DIS3L have recently been characterized after purification from human cell cultures.[16] Both were found to be exoribonucleases with properties typical of the RNR family. However, only DIS3 exhibited endonuclease activity like yeast Dis3. This is consistent with Dis3l proteins lacking two of the amino acid residues that were shown to be important for catalysis in PIN domains of other proteins.

Active Sites

The active site of the RNB domain is formed by a short stretch of the polypeptide chain containing four aspartic acid residues that coordinate magnesium ions. In all RNR proteins tested so far, exoribonuclease activity can be abolished by a single point mutation targeting one of those residues. This mutation is D209N in RNase II,[35] D280N in RNase R,[36] D551N in yeast Dis3,[12] and for human proteins: D487N in DIS3 and D486N in DIS3L[15,16] (Fig. 2).

Endoribonuclease activity of the Dis3 PIN domain is dependent on coordination of manganese ions by four acidic amino acid residues and mutation of any of these residues results in an inactive protein. These residues are D91, E120, D171, D198 for yeast Dis3,[19,33,34] and in human D69, E98, D146, D177 for DIS3 and D62, A91, T140, D166 for DIS3L.[16] Notably two of the residues in DIS3L are naturally non-acidic and consequently DIS3L does not have endoribonuclease activity (Fig. 2).

The exoribonuclease activity of the RNB domain and the endoribonuclease activity of the PIN domain are independent of each other in that the disruption of one does not affect the other. This was shown by testing mutant proteins and is also apparent from the properties of DIS3L.[16,19,33,34]

Rrp6

Rrp6 proteins are not as strongly conserved as Dis3. They are related to *Escherichia coli* RNase D, but again, the similarity is not as strong as within the eukaryotic and bacterial branches of the RNR superfamily. RNase D is ubiquitous in eukaryotes, but very rare in bacteria and archaeans lack the enzyme altogether. In fact, it has been suggested that bacteria have acquired RNase D by horizontal gene transfer from eukaryotes. RNase D and Rrp6 belong to the DEDDy family, a subset of the large DEDD superfamily, which contains various nucleases, including the nuclease domain of DNA polymerases and oligoribonuclease. The hallmark of the DEDD family is four acidic amino acid residues (three aspartate and one glutamate) required for catalysis, which is dependent on divalent metal cations. The DEDD residues do not form a continuous motif, but are distributed between three separate motifs.[2]

The domain structure of yeast Rrp6 entails the catalytic DEDD domain with the four conserved residues and one HRDC domain of unknown function. Notably, these two domains cover only one-third of the polypeptide chain and little is known about the rest of the protein. The crystal structure of Rrp6 shows the conserved amino acid residues, D238, E240, D296, D365 and Y361 coordinating two ions, one zinc and one manganese.[37]

Yeast Rrp6 has a strong tendency to aggregate and is difficult to work with, which may explain why numerous attempts at purification and thorough characterization have failed. Purification of the free protein from yeast is additionally complicated by its very high affinity to the exosome ring[12] and RNA (Andrzej Dziembowski, unpublished data). The few studies that succeeded revealed that the protein releases nucleoside 5′-monophosphates in a distributive manner. It is active only on 3′ ends with hydroxyl groups, but not on ones with phosphate groups and is unable to process structured RNAs. End products are oligonucleotides of 4-5 nt. Human Rrp6 has similar properties to the yeast protein.[10,14]

HOLOENZYME ACTIVITIES

Dis3 Exonuclease

The first studies concerning activity of the holoexosome involved purification of the native complex from *Saccharomyces cerevisiae*. This was achieved by copurification with tagged versions of either Dis3 itself or exosome ring proteins used as bait.[12,22] More or less at the same time the yeast and human exosome complexes were reconstituted in vitro. Ring subunits were expressed in *Escherichia coli*, individually or in pairs, purified and then mixed. Dis3 and Rrp6 were also expressed in bacteria, purified separately and added to the reconstituted exosome ring to obtain 10- and 11-subunit versions of the complex.[10]

In the first study demonstrating holoenzyme activity[12] the native ring+Dis3 complex had essentially the same activity on both unstructured and structured substrates, a synthetic oligonucleotide and an in vitro transcribed tRNA precursor, respectively. Differences were observed but were very modest and these activities were very similar to those observed for Dis3 alone. Later studies, however, showed that the activity of yeast Dis3 on structured substrates containing duplexes 17 bp long was inhibited by the exosome ring.[22] A different result was obtained even though the same purification procedure was applied. The reason for this inconsistency is discussed below.

For the proteins purified from bacteria, exonuclease activity of yeast Dis3 on single-stranded substrates was slightly attenuated by the ring complex, but apart from slowing down, substrate specificity and extent of degradation were unaffected. However, when structured substrates were applied, hardly any activity of the ring+Dis3 complex was observed.[10] On the other hand, in a later study such a complex did degrade structured substrates, but required a significantly longer single-stranded stretch at the 3′ end to do so.[11] These observations are readily explained by RNase protection assays and structural data. The unstructured 3′ end that Dis3 requires in order to initiate degradation is threaded into the channel of the RNB domain and reaches the catalytic site. Only when hydrolysis occurs, can secondary structures be unwound. For Dis3 alone an unpaired stretch of 10 nt is perfectly sufficient.[10,22] When Dis3 is bound to the exosome ring, the substrate must first pass through the ring channel before it can enter the RNB domain, so a longer unstructured stretch is necessary: the 10-subunit complex is inactive on substrates with a 10 nt overhang, but completely digests those with a 35 nt overhang.[11] See Figure 4 and Chapter 2 by K. Januszyk and C. Lima for more details.

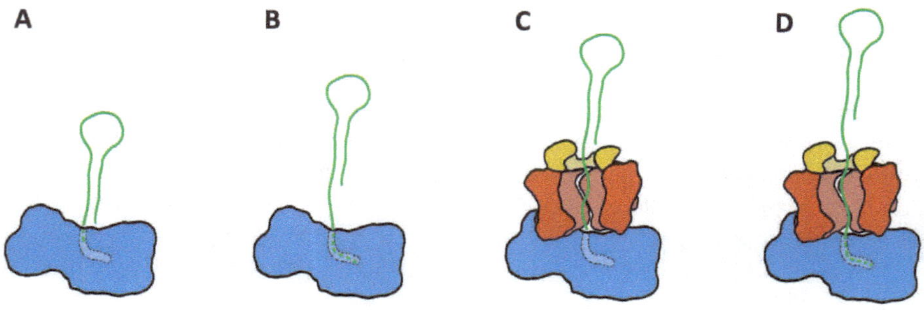

Figure 4. Substrate filtering by the exosome ring complex. Any RNA molecule with secondary structure can be digested by Dis3, provided it has a single stranded stretch at the 3′ end long enough to reach the active site buried within the RNB domain. A) Molecules with free ends shorter than 4 nt are resistant to hydrolysis. B) Molecules with free ends of 5 nt and longer can be digested to end products by free Dis3. C) Substrates threaded through the exosome ring require a longer free end to reach the active site of Dis3 and thus molecules with single-stranded 3′ ends as long as 30 nt are protected. D) Free ends longer than 33 nt are sufficient to reach the Dis3 active site through the exosome ring.

Conflicting Results with Structured Substrates

The exosome ring does not inhibit Dis3 activity on structured substrates as such, but simply changes the minimal length of the 3′ overhang that can be bound and digested (Fig. 4). Thus it is puzzling why the first study testing holoexosome activity showed that binding to the exosome ring had no significant effect on the activity of Dis3 against structured substrates,[12] especially since a very strong effect was observed in later studies.[22]

One possible explanation is that the two studies used different structured substrates: an in vitro transcribed pretRNA was degraded with similar efficiency by yeast Dis3 alone and the 10-subunit complex, whereas an intermolecular duplex with 3′ overhangs of 14 nt and less was only degraded by free Dis3. Both these substrates contain secondary structures, but with arguably different stabilities. Notably, the efficiency of hydrolysis by Dis3 is to a certain extent dependent on the length of the free 3′ end: highest activity is observed with overhangs of 10 nt and longer, but substrates with overhangs as short as 5 nt can still be degraded.[22] This is because the protein requires a single-stranded binding site in order to thread the 3′ terminus into the active site, which is also true for the exosome complex.[11] It is easy to imagine how an unstable secondary structure constitutes a smaller obstacle to binding, thermal breathing events transiently elongating the single-stranded region.

The duplex used in the second study contained a continuous 17 bp region, making for great stability. The pretRNA from the first study on the other hand was completely unmodified, which could locally weaken secondary structures and allow for efficient binding by both Dis3 and the complex. Such an effect would be in line with the ability of yeast Dis3 to recognize hypomodified tRNAs, presumably due to their misfolding.[38]

Rrp6 and Dis3 Endonuclease

Endonuclease activity of yeast Dis3 was not assayed in the study that first demonstrated complex reconstitution, as it was unknown at the time. It was later shown that binding of the protein to the ring complex has little or no effect on the activity. This suggests that RNA can access the PIN active site directly, without going through the ring channel.[11]

Activity of yeast Rrp6 was also unaffected by binding to the ring. Although activity of a ring+Rrp6 complex was never tested directly, assays were performed for Rrp6, ring+Dis3 and ring+Dis3+Rrp6. Degradation patterns for the 11-subunit complex were the sum of the patterns obtained with Rrp6 alone and with the 10-subunit complex, revealing no modulation of Rrp6 activity.[10]

Cooperation of the Three Activities

Rrp6 and Dis3 can bind to the ring concomitantly, as evidenced by copurification of the two proteins[3,12] and successful reconstitution of the 11-subunit complex.[10] The three activities of the catalytic subunits can be brought together in one molecular assembly and possibly act on the same substrate, but whether the cell prefers 11- to 10-subunit complexes (ring+Dis3 or ring+Rrp6) is unknown. Certainly in yeast Dis3 and Rrp6 can act at different steps of one process, as exemplified by processing of the 5.8S rRNA. A precursor molecule is first processed by Dis3, producing an intermediate elongated by 30 nt at the 3′ end,[39] which is then trimmed by Rrp6, leaving an 8 nt extention, later removed by the Rex RNases.[40] The length of the tail in the intermediate left by Dis3 is perfectly rationalized by threading of substrates through the exosome ring and into Dis3, although why the complex dissociates is not explained.[11] Processing of precursor RNAs first by processive and then by distributive RNases has also been suggested in tRNA maturation in bacteria[41,42] and is probably a general mechanism of RNA maturation, preventing overzealous processive enzymes from damaging the body of mature molecules.[41]

Sequential cooperative action of the endo- and exonuclease activities may also enhance the degradation potential of the exosome, allowing it to destroy troublesome molecules. It was suggested that the endoribonuclease activity of Dis3 serves to create additional entry points in substrates that cause the exosome to stall.[33] This hypothesis is strongly supported by the fact that a yeast mutant devoid of the exoribonuclease activity (*dis3*^{D551N}) accumulates decay intermediates resulting from structured molecules having been cut in predicted single-stranded regions. The same type of cleavage was observed in vitro.[19,33]

NONCATALYTIC ACTIVITIES OF THE EXOSOME RING

The exosome ring has only a modest effect on the enzymatic properties of the catalytic subunits, raising questions about the function of the ring proteins in fungi and animals. So far three explanations have been proposed for the continued existence of the ring even without catalytic functions: substrate recruitment, substrate restriction and interaction with regulatory factors.

Substrate Recruitment

The ring is considered too ancient and important to remove. The consensus is that an exosome ring-like molecular machine, a "proto-exosome" so to speak, arose even before differentiation of the three domains of life. Pathways of RNA processing and decay already existed and all were centered around a capped hexameric ring of phosphorolytic RNases. Later the system diverged into the ones we observe today: PNPase and the archaeal and eukaryotic exosomes. They do have different additional proteins and biochemical activities, but are still major knots in the RNA degradation frameworks of their respective cell systems. In bacteria and eukaryotes hydrolytic activities have been attached to the proto-exosome, both endo- and exonuclease and even though analogous proteins have not been identified in archaeans, their existence can hardly be excluded. All the while the proto-exosome has been a crucial meeting point for RNA decay substrates and effectors. In eukaryotes it lost its catalytic activity, but its other functions were too complex to be easily taken over by active enzymes. The system became trapped with an inactive assembly of proteins that are still required for bringing together enzymes, their substrates and regulatory factors.[12]

Exosome Activation

Concordantly, the exosome can and in vivo may need to, be activated by the action of additional cofactors, such as the TRAMP complex in the nucleus, which marks RNAs for exosomal destruction by polyadenylation,[43,44] the Ski complex in the cytoplasm[45] and possibly even some AU-rich element binding proteins (AUBPs).[29,30] Both TRAMP and Ski complexes contain RNA helicases[44,46] and the AUBP RHAU was also shown to bear helicase activity that is necessary for exosome activation by that protein.[47] Moreover, the Ski complex, as well as some AUBPs, have been shown to interact with the exosome ring,[29,45] and even a direct interaction of the human Rrp45 with AU-rich elements was reported.[30] Thus, the function of the ring as an interaction platform seems relatively well documented.

Substrate Filtering

It is clear that the presence of the exosome ring does not alter the activity of Dis3, but rather acts as a substrate filter, rejecting substrates with free 3′ ends too short to reach the catalytic site. This observation sheds a light on how the exosome works in the cell. It was suggested on several occasions that RNAs are targeted to degradation when their structure is compromised.[48] Be it due to a transcription error or hypomodification, deviation from the proper structure would be recognized and the molecule would be removed. Weakening the structure might expose a free 3′ end long enough to be bound by Dis3, which was indeed observed: yeast Dis3 preferentially binds to a hypomethylated $tRNA_i$ when presented with total RNA.[38] Still, since Dis3 needs a very short overhang to bind a molecule, such activity would put many normal RNAs in danger of becoming falsely identified as misfolded. Association of Dis3 with the exosome ring prevents the enzyme from attacking RNAs unless they have a long unstructured stretch (unbound by proteins), filtering potential substrates. At the same time the ring provides a way for a substrate RNA to be tagged for degradation by adding an unstructured tail, such as in the case of polyadenylation mediated by the TRAMP complex.[43,44]

This hypothesis also explains why the exosome is activated by complexes containing helicase activity. If normal, undamaged RNAs must be removed, the protective effect of their "healthy" structure and bound proteins may need to be counteracted. In vitro data suggest that Dis3 and the holoexosome can degrade quite stable secondary structures, but this was never tested on large molecules such as rRNA. Neither is there any indication of how quickly marked RNAs should be removed, but intuitively the process should be rapid and efficient. Attaching an RNA helicase activity to the exosome would speed up the process, preventing both accumulation of marked RNAs and sequestering of the exosome.

The situation of Dis3 in the complex is such that the only entrance to the RNB domain leads through the ring channel, but the active site of the PIN domain is exposed outward and freely accessible. Rrp6 is positioned on top of the ring and appears to be accessible at all times too.[49] So far regulation of these activities by the ring has not been observed.[10,11] For them the function of the ring seems to be limited to substrate recruitment and activation.

Why the exosome would carefully restrict access to one of its active sites, but permit access to two others is difficult to explain. Though perhaps no explanation is needed. It is very unlikely that such a central piece of the cell machinery would be left unchecked. It makes sense for the most aggressive activity (Dis3 exonuclease) to be the most conspicuously restrained and its control mechanism was relatively easily unraveled. The mechanisms controlling the other two activities surely exist, but are for the time being unknown.

CONCLUSION

The data reviewed above indicate that the eukaryotic exosome is a multiprotein complex endowed with ribonuclease activity. Its key structural feature is a nine-subunit entity here called the exosome ring. This ancient structure, conserved in Archaea and with analogues among Bacteria, has no catalytic activity in yeast and human, but is likely to be a phosphorolytic enzyme in its own right in plants. Despite being inactive, the ring is far from useless. It mediates interaction with some exosome substrates as well as with proteins and protein complexes that can activate the exosome. It also protects some molecules, which could be recognized as exosome substrates from undue degradation. Thus, while not necessary for catalysis itself, the ring is vital for directing the catalytic activity of the complex.

The ribonuclease activities of the exosome are provided by additional subunits: Dis3, bearing both endonuclease and processive 3'-5' exonuclease activities, and Rrp6, a distributive 3'-5' exonuclease; both proteins are hydrolases. In vertebrates and plants variants of Dis3 and Rrp6 exist. Existing data give some hints that they do not have identical activities or subcellular locations. Further work will be required to understand how the exosome composition, activity and localization is controlled in multicellular organisms.

It has now become evident that various classes of RNA are important elements of life at the molecular level, contributing to cells' normal behaviour and to their reactions to environmental challenges. As the exosome exerts its influence on most of these RNAs and has been shown to act in virtually all facets of RNA decay, both homeostatic and regulated, its significance can hardly be underestimated. It is known that the different activities of the exosome can contribute to the same physiological processes. Likewise, activities of Rrp6 and Dis3 have been shown to have their own functions. How they are coordinated and in fact whether they are coordinated at all, remain unknown.

REFERENCES

1. Mitchell P, Petfalski E, Shevchenko A et al. The exosome: a conserved eukaryotic RNA processing complex containing multiple 3'—>5' exoribonucleases. Cell 1997; 91(4):457-466.
2. Zuo Y, Deutscher MP. Exoribonuclease superfamilies: structural analysis and phylogenetic distribution. Nucleic Acids Res 2001; 29(5):1017-1026.
3. Allmang C, Petfalski E, Podtelejnikov A et al. The yeast exosome and human PM-Scl are related complexes of 3' —> 5' exonucleases. Genes Dev 1999; 13(16):2148-2158.
4. Koonin EV, Wolf YI, Aravind L. Prediction of the archaeal exosome and its connections with the proteasome and the translation and transcription machineries by a comparative-genomic approach. Genome Res2001; 11(2):240-252.
5. Symmons MF, Jones GH, Luisi BF. A duplicated fold is the structural basis for polynucleotide phosphorylase catalytic activity, processivity and regulation. Structure 2000; 8(11):1215-1226.
6. Aloy P, Ciccarelli FD, Leutwein C et al. A complex prediction: three-dimensional model of the yeast exosome. EMBO Reports 2002; 3(7):628-635.
7. Buttner K, Wenig K, Hopfner KP. Structural framework for the mechanism of archaeal exosomes in RNA processing. Mol Cell 2005; 20(3):461-471.
8. Hernandez H, Dziembowski A, Taverner T et al. Subunit architecture of multimeric complexes isolated directly from cells. EMBO Reports 2006; 7(6):605-610.
9. Lorentzen E, Walter P, Fribourg S et al. The archaeal exosome core is a hexameric ring structure with three catalytic subunits. Nat Struct Mol Biol 2005; 12(7):575-581.
10. Liu Q, Greimann JC, Lima CD. Reconstitution, activities and structure of the eukaryotic RNA exosome. Cell 2006; 127(6):1223-1237.

11. Bonneau F, Basquin J, Ebert J et al. The yeast exosome functions as a macromolecular cage to channel RNA substrates for degradation. Cell 2009; 139(3):547-559.
12. Dziembowski A, Lorentzen E, Conti E et al. A single subunit, Dis3, is essentially responsible for yeast exosome core activity. Nat Struct Mol Biol 2007; 14(1):15-22.
13. Chekanova JA, Shaw RJ, Wills MA et al. Poly(A) tail-dependent exonuclease AtRrp41p from Arabidopsis thaliana rescues 5.8 S rRNA processing and mRNA decay defects of the yeast ski6 mutant and is found in an exosome-sized complex in plant and yeast cells. J Biol Chem 2000; 275(42):33158-33166.
14. Burkard KT, Butler JS. A nuclear 3′-5′ exonuclease involved in mRNA degradation interacts with Poly(A) polymerase and the hnRNA protein Npl3p. Mol Cell Biol 2000; 20(2):604-616.
15. Staals RHJ, Bronkhorst AW, Schilders G et al. Dis3-like 1: a novel exoribonuclease associated with the human exosome. submitted.
16. Tomecki R, Kristiansen MS, Lykke-Andersen S et al. The human core exosome interacts with differentially localized processive ribonucleases: hDIS3 and hDIS3L. EMBO J, accepted manuscript.
17. Lange H, Holec S, Cognat V et al. Degradation of a polyadenylated rRNA maturation by-product involves one of the three RRP6-like proteins in Arabidopsis thaliana. Mol Cell Biol 2008; 28(9):3038-3044.
18. Chekanova JA, Gregory BD, Reverdatto SV et al. Genome-wide high-resolution mapping of exosome substrates reveals hidden features in the Arabidopsis transcriptome. Cell 2007; 131(7):1340-1353.
19. Schneider C, Leung E, Brown J et al. The N-terminal PIN domain of the exosome subunit Rrp44 harbors endonuclease activity and tethers Rrp44 to the yeast core exosome. Nucleic Acids Res 2009; 37(4):1127-1140.
20. Cheng ZF, Deutscher MP. Purification and characterization of the Escherichia coli exoribonuclease RNase R. Comparison with RNase II. J Biol Chem 2002; 277(24):21624-21629.
21. Frazao C, McVey CE, Amblar M et al. Unravelling the dynamics of RNA degradation by ribonuclease II and its RNA-bound complex. Nature 2006; 443(7107):110-114.
22. Lorentzen E, Basquin J, Tomecki R et al. Structure of the active subunit of the yeast exosome core, Rrp44: diverse modes of substrate recruitment in the RNase II nuclease family. Mol Cell 2008; 29(6):717-728.
23. Makarova KS, Aravind L, Galperin MY et al. Comparative genomics of the Archaea (Euryarchaeota): evolution of conserved protein families, the stable core and the variable shell. Genome Res 1999; 9(7):608-628.
24. Arcus VL, Backbro K, Roos A et al. Distant structural homology leads to the functional characterization of an archaeal PIN domain as an exonuclease. J Biol Chem 2004; 279(16):16471-16478.
25. Clissold PM, Ponting CP. PIN domains in nonsense-mediated mRNA decay and RNAi. Curr Biol 2000; 10(24):R888-890.
26. Glavan F, Behm-Ansmant I, Izaurralde E et al. Structures of the PIN domains of SMG6 and SMG5 reveal a nuclease within the mRNA surveillance complex. EMBO J 2006; 25(21):5117-5125.
27. Mamolen M, Andrulis ED. Characterization of the Drosophila melanogaster Dis3 ribonuclease. Biochem Biophys Res Commun 2009; 390(3):529-534.
28. Gunther T. Concentration, compartmentation and metabolic function of intracellular free Mg2+. Magnes Res 2006; 19(4):225-236.
29. Chen CY, Gherzi R, Ong SE et al. AU binding proteins recruit the exosome to degrade ARE-containing mRNAs. Cell 2001; 107(4):451-464.
30. Mukherjee D, Gao M, O'Connor JP et al. The mammalian exosome mediates the efficient degradation of mRNAs that contain AU-rich elements. EMBO J 2002; 21(1-2):165-174.
31. Vincent HA, Deutscher MP. The roles of individual domains of RNase R in substrate binding and exoribonuclease activity. The nuclease domain is sufficient for digestion of structured RNA. J Biol Chem 2009; 284(1):486-494.
32. Vincent HA, Deutscher MP. Insights into how RNase R degrades structured RNA: analysis of the nuclease domain. J Mol Biol 2009; 387(3):570-583.
33. Lebreton A, Tomecki R, Dziembowski A et al. Endonucleolytic RNA cleavage by a eukaryotic exosome. Nature 2008; 456(7224):993-996.
34. Schaeffer D, Tsanova B, Barbas A et al. The exosome contains domains with specific endoribonuclease, exoribonuclease and cytoplasmic mRNA decay activities. Nat Struct Mol Biol 2009; 16(1):56-62.
35. Amblar M, Arraiano CM. A single mutation in Escherichia coli ribonuclease II inactivates the enzyme without affecting RNA binding. FEBS J 2005; 272(2):363-374.
36. Matos RG, Barbas A, Arraiano CM. RNase R mutants elucidate the catalysis of structured RNA: RNA-binding domains select the RNAs targeted for degradation. Biochem J 2009; 423(2):291-301.
37. Midtgaard SF, Assenholt J, Jonstrup AT et al. Structure of the nuclear exosome component Rrp6p reveals an interplay between the active site and the HRDC domain. Proc Natl Acad Sci USA 2006; 103(32):11898-11903.
38. Schneider C, Anderson JT, Tollervey D. The exosome subunit Rrp44 plays a direct role in RNA substrate recognition. Mol Cell 2007; 27(2):324-331.

39. Briggs MW, Burkard KT, Butler JS. Rrp6p, the yeast homologue of the human PM-Scl 100-kDa autoantigen, is essential for efficient 5.8 S rRNA 3' end formation. J Biol Chem 1998; 273(21):13255-13263.
40. van Hoof A, Lennertz P, Parker R. Three conserved members of the RNase D family have unique and overlapping functions in the processing of 5S, 5.8S, U4, U5, RNase MRP and RNase P RNAs in yeast. EMBO J 2000; 19(6):1357-1365.
41. Deutscher MP. Degradation of RNA in bacteria: comparison of mRNA and stable RNA. Nucleic Acids Res 2006; 34(2):659-666.
42. Li Z, Pandit S, Deutscher MP. 3' exoribonucleolytic trimming is a common feature of the maturation of small, stable RNAs in Escherichia coli.Proc Natl Acad Sci USA 1998; 95(6):2856-2861.
43. LaCava J, Houseley J, Saveanu C et al. RNA degradation by the exosome is promoted by a nuclear polyadenylation complex. Cell 2005; 121(5):713-724.
44. Vanacova S, Wolf J, Martin G et al. A new yeast poly(A) polymerase complex involved in RNA quality control. PLoS Biol 2005; 3(6):e189.
45. Araki Y, Takahashi S, Kobayashi T et al. Ski7p G protein interacts with the exosome and the Ski complex for 3'-to-5' mRNA decay in yeast. EMBO J 2001; 20(17):4684-4693.
46. Brown JT, Bai X, Johnson AW. The yeast antiviral proteins Ski2p, Ski3p and Ski8p exist as a complex in vivo. RNA (New York, N.Y.) 2000; 6(3):449-457.
47. Tran H, Schilling M, Wirbelauer C et al. Facilitation of mRNA deadenylation and decay by the exosome-bound, DExH protein RHAU. Mol Cell 2004; 13(1):101-111.
48. Reinisch KM, Wolin SL. Emerging themes in noncoding RNA quality control. Curr Opin Struct Biol 2007; 17(2):209-214.
49. Cristodero M, Bottcher B, Diepholz M et al. The Leishmania tarentolae exosome: purification and structural analysis by electron microscopy. Mol Biochem Parasitol 2008; 159(1):24-29.

CHAPTER 7

FUNCTIONS OF THE CYTOPLASMIC EXOSOME

Daneen Schaeffer, Amanda Clark,[†] A. Alejandra Klauer,[†] Borislava Tsanova[†] and Ambro van Hoof*

Department of Microbiology and Molecular Genetics, University of Texas Health Science Center-Houston, Houston, Texas, USA.

[†]These authors contributed equally.

**Corresponding Author: Ambro van Hoof—Email: ambro.van.hoof@uth.tmc.edu*

Abstract: The exosome consists of a core of ten essential proteins that includes the ribonuclease Rrp44p and is present in both the cytoplasm and nucleus of eukaryotic cells. The cytoplasmic exosome has been extensively characterized in the budding yeast *Saccharomyces cerevisiae* and some characterization of its metazoan counterpart indicates that most functional aspects are conserved. These studies have implicated the cytoplasmic exosome in the turnover of normal cellular mRNAs, as well as several mRNA surveillance pathways. For this, the exosome needs a set of four proteins that do not partake in nuclear exosome functions. These cofactors presumably direct the exosome to specific cytoplasmic RNA substrates. Here, we review cofactors and functions of the cytoplasmic exosome and provide unanswered questions on the mechanisms of cytoplasmic exosome function.

INTRODUCTION

The eukaryotic exosome is present in both the nucleus and cytoplasm and carries out a variety of RNA processing and degradation reactions. The nuclear and cytoplasmic forms of the exosome contain the same 10 essential subunits as reviewed in other chapters in this book. Also described in other chapters in this book are the various RNA processing and degradation events catalyzed by the nuclear exosome, with the help of a number of nuclear cofactors. In this chapter we will focus on the cytoplasmic roles of the exosome and on four cofactors solely required for cytoplasmic functions of the exosome.

RNA Exosome, edited by Torben Heick Jensen.

THE CYTOPLASMIC EXOSOME REQUIRES FOUR COFACTORS

Three genetic screens have been performed that identified cofactors of the cytoplasmic exosome. Two screens for host mutants that affect the killer virus system (see below) identified exosome subunits as well as four proteins named superkiller, or Ski, proteins.[1,2] These four Ski proteins were later shown to be cytoplasmic exosome cofactors.[3,4] The cytoplasmic exosome is also required for the rapid degradation of nonstop mRNAs (see below). A genetic screen of the genome-wide collection of yeast knock-outs for mutants defective in nonstop mRNA decay identified the same four *SKI* genes.[5] In this nonstop mRNA decay screen, a number of other mutants also showed smaller effects on the expression of nonstop mRNAs, but these additional mutations do not appear to be in exosome cofactors. Since the killer virus screen identified many alleles of each of the four *SKI* genes and the genome-wide nonstop decay screen identified the same four genes, it appears that there are only four cofactors of the cytoplasmic exosome. Various experiments have shown that these four Ski proteins are needed for all known functions of the cytoplasmic exosome.[3,6-8] This differs from the nuclear exosome cofactors, which are more numerous and mostly appear to be required for only a subset of exosome functions (with the exception of Mtr4p). An additional difference is that the cytoplasmic exosome cofactors are not required for viability, while many nuclear exosome cofactors and the exosome subunits themselves are essential. This indicates that the activity of the cytoplasmic exosome is not essential, presumably because its role overlaps with that of Xrn1p (see below). Since deletion of the SKI genes does not appear to affect the distribution of the exosome between the nucleus and cytoplasm (A. van Hoof and Roy Parker; unpublished data), the Ski proteins are most likely directly involved in exosome function.

The Ski Complex

Three of the Ski proteins form a Ski complex, consisting of Ski2p, Ski3p and two copies of Ski8p.[9,10] Ski2p is the only Ski protein with a catalytic function, while Ski3p and Ski8p contain motifs thought to be needed for protein-protein interactions. Ski2p is a putative DExH-box RNA helicase and mutating the DEVH motif to AEVA disrupts Ski2p function, suggesting that Ski2p is catalytically active and that this activity is required for its function.[11] Ski2p contains 1287 amino acids and the C-terminal 960 amino acids resemble the nuclear exosome cofactor Mtr4p. The structure of Mtr4p was recently solved and revealed four domains that are shared with a family of eukaryotic and archaeal RNA and DNA helicases (the Ski2-like family of DExH helicases) and thus form a helicase core.[12] In addition, a fifth domain shared between Mtr4p and Ski2p has been termed the arch domain, because it forms an arch-like structure on one side of the helicase core.[12] This arch domain is present in all eukaryotic Ski2p and Mtr4p orthologs, but not in any other protein, distinguishing exosome-associated RNA helicases from other helicases. Although the Mtr4p arch domain is required for exosome-mediated RNA decay and processing, its molecular functions and the function of the Ski2p arch, are not yet clear.[12] Mtr4p and Ski2p also contain a large conserved surface area, which has been suggested to interact with the conserved surface of the cap proteins of the exosome.[12] The N-terminal 320 residues of Ski2p do not show any sequence similarity with Mtr4p and likely provide functions required for Ski2p, but not Mtr4p. Consistent with this, Wang et al.[11] used yeast two-hybrid and co-immunoprecipitation experiments to show that this region of Ski2p is necessary and sufficient for interaction with Ski3p and Ski8p. Therefore, the Ski complex seems to have a general helicase core and two accessory domains. The first accessory domain is an arch domain that is shared only with exosome-associated RNA helicases, while the second accessory domain is a Ski complex-specific part and consists of Ski3p, two copies of Ski8p and the N-terminus of Ski2p.

Although the function of cytoplasmic exosome cofactors has not been extensively studied in other eukaryotes, such as humans, the available evidence suggests that the function of all four is conserved in Metazoans. The *Drosophila* Ski2p (a.k.a. twister), Ski3p and Ski8p are required for cytoplasmic 3′ to 5′ mRNA degradation and thus carry out essentially the same function as the yeast orthologs.[13-15]

Most eukaryotes, including humans, have two RNA helicases of the Mtr4p/Ski2p group with one more closely resembling Ski2p and the other more closely resembling Mtr4p (unpublished observations), however the very early diverging genera *Trypanosoma*, *Leishmania* and *Giardia* have only one recognizable ortholog that is most similar in sequence to Mtr4p (unpublished observations and ref. 16). It is not entirely clear whether, in these species, one protein carries out both the Mtr4p and Ski2p function, or whether they only need Mtr4p or Ski2p function. The *T. brucei* protein resembles Mtr4p in that when overexpressed and tagged it is mainly nuclear,[16] it forms a TRAMP complex[17] and knocking down its expression causes a defect in rRNA processing.[16] In addition, *T. Brucei* lacks readily identifiable orthologs of Ski3p and Ski8p. Thus, while most eukaryotes have a Ski complex, *T. brucei* may lack Ski complex function. Interestingly, the exosome in *T. brucei* is mostly cytoplasmic[18] and thus may function independently of a Ski complex.

Ski7p

In addition to the Ski complex, the cytoplasmic exosome also requires Ski7p, which is composed of two functional domains. The N-terminus of Ski7p interacts with the cytoplasmic exosome and the Ski complex[19] and is needed for exosome-mediated degradation of both normal and nonstop mRNAs.[8,19] Interestingly, this region appears unique to Ski7p orthologs and is not similar to any known protein. The C-terminal domain of Ski7p is homologous to the translation factors eEF1A and eRF3[4,19,20] and is important in nonstop mRNA decay (see below; ref. 8).

While the evolutionary history of the Ski complex is straightforward, the yeast *SKI7* gene has a peculiar history. About 100 million years ago, the ancestor of yeast duplicated its genome. Subsequently, most of the duplicated genes were lost, but both *SKI7* and its paralog *HBS1* were maintained.[21,22] As a consequence of this duplication, *S. cerevisiae* and its close relatives have both a *SKI7* gene and an *HBS1* gene, but most other eukaryotes have only one corresponding gene that presumably performs the functions of both *SKI7* and *HBS1*. Consistent with this, the single *S. kluyveri* gene can perform both *SKI7* and *HBS1* function.[23] In addition, knocking down the expression of the *Drosophila HBS1/SKI7* ortholog inhibits cytoplasmic exosome function, although to a lesser extent than knocking down a component of the Ski complex (leading to a 2-fold increase in the abundance of an mRNA decay intermediate, compared to a 10-fold increase after Ski complex knockdown; ref. 14).

In conclusion, most, but not all, eukaryotes have orthologs of the four *SKI* genes identified in yeast and these orthologs presumably carry out the same functions. The function of the yeast *SKI* genes has been characterized using genetics and based on these analyses, the Ski complex and Ski7p are thought to recruit the cytoplasmic exosome to specific RNA substrates. However, their biochemistry and how they act on the molecular level is still largely unexplored. The structures of Mtr4p[12] and Ski8p[24] and future biochemical approaches should be very useful in increasing our understanding of the function of the cytoplasmic exosome cofactors.

THE EXOSOME FUNCTIONS IN ONE OF TWO GENERAL PATHWAYS FOR CYTOPLASMIC mRNA DEGRADATION

Two general pathways of mRNA degradation have been identified using *S. cerevisiae* as a model system and both pathways appear conserved in most other eukaryotes (Fig. 1). The initiating and rate-limiting step in both pathways is the shortening of the poly(A) tail in a process termed deadenylation. Following deadenylation, transcripts can be degraded from their 5′- and 3′-ends. The 5′ to 3′ mRNA decay pathway is initiated by removal of the 5′ cap by Dcp2p.[25-27] This exposes the 5′-end of an mRNA to degradation by the

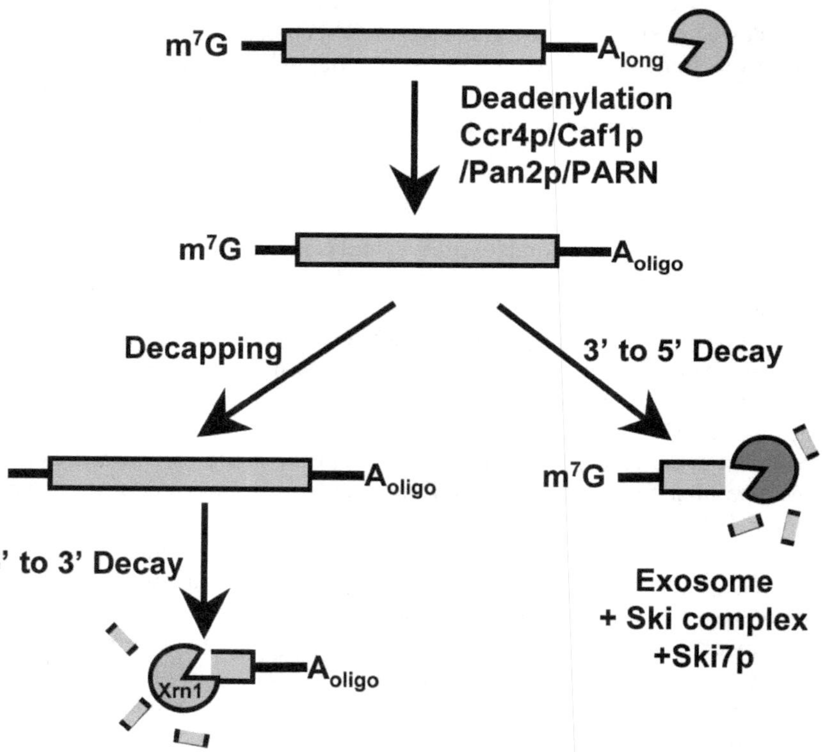

Figure 1. The degradation of eukaryotic mRNAs is generally initiated by removal of the poly(A) tail, which can be carried out by various deadenylases (Ccr4p, Caf1p, Pan2p and/or PARN). Deadenylation can be followed by removal of the cap structure, which makes the RNA susceptible to the 5′ to 3′ exoribonuclease Xrn1p. Alternatively, deadenylation can be followed by 3′ to 5′ degradation by the exoribonuclease activity of the cytoplasmic exosome. This activity of the exosome also requires Ski7p and the Ski complex.

5′ to 3′ exoribonuclease Xrn1p.[12,28-30] Deadenylation can also trigger the degradation of an mRNA by the cytoplasmic exosome.[3] Following exosome-mediated decay of the body of the transcript, the scavenger decapping enzyme, Dcs1p in yeast and DcpS in humans, removes the 5′ cap from the remaining oligonucleotide.[31,32]

These two general decay pathways are redundant and mRNAs can be degraded either by the decapping or the exosome pathway. Consistent with this, neither pathway is essential for yeast viability, but simultaneous inactivation of both mRNA degradation pathways is lethal.[3,33] Although mRNAs can be degraded by either pathway, normal yeast mRNAs are mostly degraded through the decapping pathway, with the exosome making a smaller contribution. This conclusion is based on the observation that several normal cellular mRNAs are stabilized in mutants of the decapping enzyme or of Xrn1p, but not in mutants defective in cytoplasmic exosome function. Careful determination of key decay characteristics suggests that exosome-mediated decay is about 2-5 fold slower than decapping-mediated decay in yeast.[34] Quantification of this difference required the use of null mutants that are unavailable in most other organisms, thus it is not clear whether this 2-5 fold rate difference is conserved, or whether there are conditions or other eukaryotic cells where the relative importance is reversed.

The exosome contains both endoribonuclease and 3′ to 5′ exoribonuclease domains[35-37] and the 3′ to 5′ exoribonuclease activity is the major contributor to the exosome's mRNA decay function. This conclusion is based on two observations. First, a mutation that inactivates the 3′ exoribonuclease domain leads to accumulation of mRNA decay intermediates (in a strain background also lacking decapping activity; ref. 38). Second, this same 3′ exoribonuclease mutation is synthetic lethal with deletion of the 5′ to 3′ exoribonuclease Xrn1p, while a mutation inactivating the endoribonuclease activity of the exosome is not.[37]

Although all mRNAs are subject to both mRNA decay pathways, individual mRNAs are degraded at different rates. For example, mammalian transcripts that contain AU-Rich Elements (ARE) in their 3′ untranslated region are more rapidly degraded.[39,40] These AREs are recognized by ARE-binding proteins, including KSRP and TTP, which also interact with the exosome to promote rapid mRNA degradation.[41-43] Thus, sequence-specific RNA binding proteins can accelerate the decay of specific mRNAs by recruiting the exosome to these mRNAs.

THE CYTOPLASMIC EXOSOME FUNCTIONS IN mRNA SURVEILLANCE

In addition to degrading "normal" cellular transcripts and mRNAs with specific sequences, the cytoplasmic exosome is also involved in the quality control of mRNAs in the cytoplasm. In these specialized mRNA surveillance pathways, exosome cofactors and adaptor proteins distinguish aberrant mRNAs from normal mRNAs and ultimately direct these aberrant transcripts to the exosome for rapid degradation. Most importantly, the cytoplasmic exosome is required for the rapid degradation of mRNAs that lack a stop codon. The cytoplasmic exosome also contributes to several other mRNA surveillance pathways, including the degradation of mRNAs with premature stop codons and mRNA fragments that are generated by various endoribonucleases. Not surprisingly, the rate-limiting step of deadenylation is bypassed in these mRNA surveillance pathways to rapidly rid the cell of these potentially harmful transcripts.

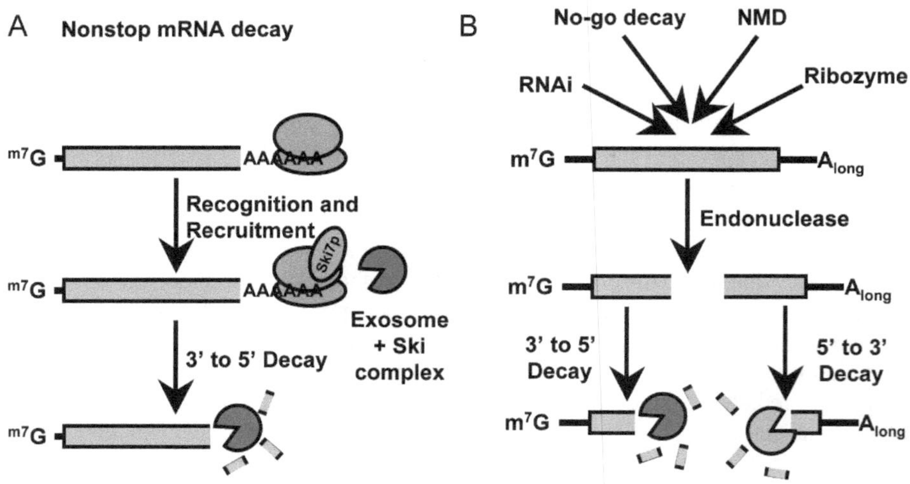

Figure 2. A) mRNAs that lack a stop codon are recognized by the C-terminal domain of Ski7p. This recruits the exosome and the Ski complex, which mediates their degradation. B) A variety of triggers can cause an mRNA to be cleaved by an endonuclease. The resulting 5′ cleavage products are degraded by the cytoplasmic exosome. This activity of the exosome also requires Ski7p and the Ski complex.

Nonstop mRNA Degradation

Mutations that inactivate the cytoplasmic exosome have little effect on the stability of normal mRNAs, but dramatically stabilize mRNAs that lack a stop codon.[8,44] These types of transcripts can arise from mistakes in gene expression, including genetic mutations, defects in transcription, or premature polyadenylation, due to inaccurate 3′ end formation or through the use of a cryptic polyadenylation site. In the current model of nonstop mRNA decay, the translating ribosome reads through the poly(A) tail and stalls at the 3′ end of the mRNA (Fig. 2A). Based on its homology to translation factors, the C-terminus of the cytoplasmic exosome cofactor Ski7p is thought to recognize the stalled ribosome with an empty A-site. Consistent with this, deletion of this C-terminal domain stabilizes nonstop mRNAs, but does not affect other exosome functions. Along with the Ski complex, Ski7p recruits the exosome to rapidly degrade the transcript from the 3′-end.[8] For a more detailed review of nonstop mRNA decay see Wilson et al.[45]

Nonsense-Mediated mRNA Degradation (NMD)

mRNAs that contain a premature nonsense codon are rapidly degraded by an mRNA surveillance pathway called nonsense-mediated mRNA degradation (NMD). How these mRNAs are distinguished from normal mRNAs is not yet clear, although the spatial relationship between the termination codon and various other features of the mRNA have been implicated.[46-49] The recognition and/or decay of these NMD targets requires a set of three Upf proteins that are conserved in most eukaryotes.[50-54] In yeast, NMD targets are predominantly degraded by decapping and 5′ to 3′ decay.[55] Nonsense transcripts are also degraded, to a lesser extent, by the cytoplasmic exosome.[7,56] The key observation supporting this conclusion is that nonsense mRNAs are more stable in *xrn1Δ upf1Δ* or *dcp1-2 upf1Δ* double mutants than in *xrn1Δ* or *dcp1-2* single mutants.[7,56]

Degradation of Endoribonuclease Products

While eukaryotes contain two general mRNA decay pathways that are mediated by exoribonucleases, in specific cases mRNA decay can be initiated by endoribonuclease cleavage (Fig. 2B). One example of mRNA decay initiated by endonucleolytic cleavage is found in the NMD pathways in *Drosophila* and humans.[13,57,58] Here, nonsense transcripts are endonucleolytically cleaved by SMG6, which generates a 5′- and 3′-degradation fragment. The 5′-fragment is degraded by the cytoplasmic exosome, in a manner that also requires the Ski complex. The 3′-degradation fragment is degraded by the 5′ to 3′ exoribonuclease Xrn1p.[13,57,58]

A second example of mRNA decay that is initiated by endonucleolytic cleavage is a process termed no-go decay. No-go decay is triggered by translational pauses caused by secondary structures such as stem loops in the mRNA, but can also be activated to a lesser extent by pseudoknots and rare codons.[6] In the current model of no-go decay, Dom34p and Hbs1p promote, but are not absolutely required for, the endonucleolytic cleavage of the transcript in the vicinity of the stalled ribosome. After cleavage, 5′ and 3′ fragments of the translational stall site are released.[6] The 5′-fragment is degraded by the exosome and this reaction also requires the Ski complex and Ski7p, while the 3′-fragment is degraded by Xrn1p.

A third example of mRNA decay that is initiated by endoribonucleolytic cleavage is RNAi. In RNAi, double-stranded RNA (dsRNA) is processed by Dicer into 21-22 nucleotide small interfering RNAs (siRNAs).[59-62] These siRNAs are then incorporated into the RNA-induced silencing complex (RISC), which can endonucleolytically cleave target transcripts at a site complementary to the siRNA.[63,64] The 5′- and 3′-fragments generated by endonucleolytic cleavage in RNAi in *Drosophila* cells are degraded by the cytoplasmic exosome and Xrn1p, respectively. The 3′ decay of the 5′-fragment also requires the Ski complex.[14]

Finally, endoribonucleolytic cleavage can be initiated by hammerhead ribozymes. This can either occur because of natural ribozyme sequences in an mRNA,[65-67] or because of artificially introduced ribozymes. Meaux and van Hoof[68] used either a hammerhead ribozyme, or a mutated group I intron to artificially generate cleaved mRNAs in yeast. The 5′ cleavage product of these reactions was degraded by the cytoplasmic exosome, with the help of cytoplasmic exosome cofactors. These 5′ cleavage products were very unstable in wild type cells (half-life 1 minute). In most of the other endoribonucleolytic cleavage reactions mentioned above, the cleavage products are not detectable unless the activity of the cytoplasmic exosome is reduced, suggesting that the degradation of these cleavage products by the exosome is also very rapid.

THE ANTIVIRAL FUNCTION OF THE CYTOPLASMIC EXOSOME

Many yeast strains contain the L-A virus and the M satellite RNA. The M satellite RNA encodes a protein toxin that is secreted from infected cells that kills uninfected cells. Thus, this toxin gives infected cells an advantage in their competition with uninfected cells for nutrients. Mutants in the *SKI* genes were discovered as "superkiller" mutants, or mutants that had an increase in viral toxin production.[2] Specifically, it was shown that disruption of any of the *SKI* genes resulted in an increase in viral copy number and increased toxin production.[1,69,70] These observations suggest that the Ski proteins have an antiviral function and regulate the replication of RNA viruses.[71] It was first hypothesized that the Ski proteins inhibit translation of nonpoly(A) viral mRNA. This hypothesis was based on the observation that only RNA that lacked a poly(A) tail were affected by *ski* mutations.[71,72] In addition, Ski7p is homologous to translation factors, suggesting that at least Ski7p has a role in translation.[20] More recent studies, however, have shown that a major function of the Ski proteins is in the decay of cytoplasmic mRNAs (see above). From these results an alternative mechanism for the Ski genes appears more likely: Instead of regulating translation, the Ski proteins target mRNAs that lack a poly(A) tail, including viral RNA, for exosome-mediated degradation. An increase in viral RNA stability would allow an increase in toxin production, thus explaining the superkiller phenotype. The M and L-A viral mRNAs resemble the endoribonuclease cleavage products discussed above, since both lack a poly(A) tail. This raises the possibility that recognition of endoribonuclease cleavage products and viral RNAs uses the same molecular mechanism. If such a common mechanism exists, it may have initially evolved as an antiviral defense, or as an mRNA surveillance pathway to correct errors in gene expression. A deeper understanding of the recognition mechanism is required to provide insight in this area.

Eukaryotes have numerous pathways that recognize viral RNAs and trigger antiviral defenses. Typical eukaryotic mRNAs are single-stranded, have a 5′ cap and a 3′ poly(A) tail. Host cells initiate innate immune defense pathways in response to RNAs that lack these specific characteristics. Examples of this include the recognition of dsRNA by toll-like receptors, by MDA5, by PKR and by 2-5A synthetase in mammalian cells (reviewed in refs. 73-75). Other eukaryotes use the RNAi machinery to recognize dsRNA and defend against viruses (e.g., refs. 76-79). The human RIG-I recognizes uncapped RNAs that end in a triphosphate, possibly with some sequence specificity.[80-82] The exosome-mediated decay of unadenylated mRNAs may similarly be considered an innate immune pathway. Whether viral mRNAs in other eukaryotes are similarly targeted is unknown.

CONCLUSION AND FUTURE PERSPECTIVES

The cytoplasmic exosome plays a dual role in eukaryotic gene expression. First, the exosome regulates gene expression by participating in general mRNA turnover to degrade transcripts that are no longer needed. Second, the exosome acts as a quality control mechanism to maintain the fidelity of gene expression by rapidly degrading aberrant transcripts. Specifically, the cytoplasmic exosome degrades transcripts that lack termination codons in the nonstop mRNA decay pathway and contributes to the degradation of transcripts that have premature termination codons or that have been cleaved. Importantly, the Ski complex and Ski7p in yeast, are needed for both general mRNA degradation and for mRNA surveillance in the cytoplasm.

Despite extensive studies of the cytoplasmic exosome in mRNA degradation, important questions remain unanswered. First, how is the exosome recruited to RNA substrates? In each of the mRNA surveillance pathways described here, one or more proteins recognize an aberrant RNA, which is then targeted for rapid degradation by the exosome. It is not yet known how these proteins and exosome cofactors recruit and/or potentially activate the nuclease activities of the exosome. Second, what is the relationship between the cytoplasmic and nuclear exosome? Does the exosome shuttle between the two compartments, or are there separate pools of nuclear and cytoplasmic exosomes? Where is the exosome assembled and how is it targeted to the nucleus, cytoplasm and nucleolus? Third, is the recently characterized endoribonuclease activity of the Rrp44p PIN domain involved in any of the functions of the cytoplasmic exosome? Fourth, are there any proteins that protect exosome substrates from degradation? All of the proteins and cytoplasmic exosome cofactors described here aid in mRNA destabilization. However the data reviewed here, do not exclude the possibility that there are proteins that stabilize cytoplasmic exosome substrates. Fifth, is the cytoplasmic exosome involved in other cytoplasmic RNA surveillance pathways? Future studies are needed to answer these questions.

ACKNOWLEDGEMENTS

Work in the authors' laboratory was supported by a grant from NIH (GM069900). We thank Drs. Ziyin Li and Jason Rosenzweig for comments on the manuscript and Dr. Roy Parker for permission to cite unpublished data.

NOTE ADDED IN PROOFS

A second paper describing the Mtr4p structure was recently published, which confirms the domain structure of Mtr4p and Ski2p.[83]

REFERENCES

1. Toh EA, Guerry P, Wickner RB. Chromosomal superkiller mutants of Saccharomyces cerevisiae. J Bacteriol 1978; 136(3):1002-7.
2. Toh EA, Wickner RB. "Superkiller" mutations suppress chromosomal mutations affecting double-stranded RNA killer plasmid replication in saccharomyces cerevisiae. Proc Natl Acad Sci USA 1980; 77(1):527-30.
3. Jacobs Anderson JS, Parker R. The 3′ to 5′ degradation of yeast mRNAs is a general mechanism for mRNA turnover that requires the SKI2 DEVH box protein and 3′ to 5′ exonucleases of the exosome complex. EMBO J 1998; 17(5):1497-506.
4. van Hoof A, Staples RR, Baker RE et al. Function of the ski4p (Csl4p) and Ski7p proteins in 3′-to-5′ degradation of mRNA. Mol Cell Biol 2000; 20(21):8230-43.
5. Wilson MA, Meaux S, van Hoof A. A genomic screen in yeast reveals novel aspects of nonstop mRNA metabolism. Genetics 2007; 177(2):773-84.
6. Doma MK, Parker R. Endonucleolytic cleavage of eukaryotic mRNAs with stalls in translation elongation. Nature 2006; 440(7083):561-4.
7. Mitchell P, Tollervey D. An NMD pathway in yeast involving accelerated deadenylation and exosome-mediated 3′—>5′ degradation. Mol Cell 2003; 11(5):1405-13.
8. van Hoof A, Frischmeyer PA, Dietz HC et al. Exosome-mediated recognition and degradation of mRNAs lacking a termination codon. Science 2002; 295(5563):2262-4.
9. Synowsky SA, Heck AJ. The yeast Ski complex is a hetero-tetramer. Protein Sci 2008; 17(1):119-25.
10. Brown JT, Bai X, Johnson AW. The yeast antiviral proteins Ski2p, Ski3p and Ski8p exist as a complex in vivo. RNA 2000; 6(3):449-57.
11. Wang L, Lewis MS, Johnson AW. Domain interactions within the Ski2/3/8 complex and between the Ski complex and Ski7p. RNA 2005; 11(8):1291-302.
12. Jackson RN, Klauer AA, Hintze BJ et al. The crystal structure of Mtr4 reveals a novel arch domain required for rRNA processing. EMBO J 2010; in press.
13. Gatfield D, Izaurralde E. Nonsense-mediated messenger RNA decay is initiated by endonucleolytic cleavage in Drosophila. Nature 2004; 429(6991):575-8.

14. Orban TI, Izaurralde E. Decay of mRNAs targeted by RISC requires XRN1, the Ski complex and the exosome. RNA 2005; 11(4):459-69.
15. Seago JE, Chernukhin IV, Newbury SF. The Drosophila gene twister, an orthologue of the yeast helicase SKI2, is differentially expressed during development. Mech Dev 2001; 106(1-2):137-41.
16. Cristodero M, Clayton CE. Trypanosome MTR4 is involved in rRNA processing. Nucleic Acids Res 2007; 35(20):7023-30.
17. Etheridge RD, Clemens DM, Gershon PD et al. Identification and characterization of nuclear noncanonical poly(A) polymerases from Trypanosoma brucei. Mol Biochem Parasitol 2009; 164(1):66-73.
18. Haile S, Cristodero M, Clayton C et al. The subcellular localisation of trypanosome RRP6 and its association with the exosome. Mol Biochem Parasitol 2007; 151(1):52-8.
19. Araki Y, Takahashi S, Kobayashi T et al. Ski7p G protein interacts with the exosome and the Ski complex for 3′- to-5′ mRNA decay in yeast. EMBO J 2001; 20(17):4684-93.
20. Benard L, Carroll K, Valle RC et al. The ski7 antiviral protein is an EF1-alpha homolog that blocks expression of nonPoly(A) mRNA in Saccharomyces cerevisiae. J Virol 1999; 73(4):2893-900.
21. Kellis M, Birren BW, Lander ES. Proof and evolutionary analysis of ancient genome duplication in the yeast Saccharomyces cerevisiae. Nature 2004; 428(6983):617-24.
22. Wolfe KH, Shields DC. Molecular evidence for an ancient duplication of the entire yeast genome. Nature 1997; 387(6634):708-13.
23. van Hoof A. Conserved functions of yeast genes support the duplication, degeneration and complementation model for gene duplication. Genetics 2005; 171(4):1455-61.
24. Cheng Z, Liu Y, Wang C et al. Crystal structure of Ski8p, a WD-repeat protein with dual roles in mRNA metabolism and meiotic recombination. Protein Sci 2004; 13(10):2673-84.
25. Dunckley T, Parker R. The DCP2 protein is required for mRNA decapping in Saccharomyces cerevisiae and contains a functional MutT motif. EMBO J 1999; 18(19):5411-22.
26. van Dijk E, Cougot N, Meyer S et al. Human Dcp2: a catalytically active mRNA decapping enzyme located in specific cytoplasmic structures. EMBO J 2002; 21(24):6915-24.
27. Steiger M, Carr-Schmid A, Schwartz DC et al. Analysis of recombinant yeast decapping enzyme. RNA 2003; 9(2):231-7.
28. Decker CJ, Parker R. A turnover pathway for both stable and unstable mRNAs in yeast: evidence for a requirement for deadenylation. Genes Dev 1993; 7(8):1632-43.
29. Hsu CL, Stevens A. Yeast cells lacking 5′—>3′ exoribonuclease 1 contain mRNA species that are poly(A) deficient and partially lack the 5′ cap structure. Mol Cell Biol 1993; 13(8):4826-35.
30. Muhlrad D, Decker CJ, Parker R. Deadenylation of the unstable mRNA encoded by the yeast MFA2 gene leads to decapping followed by 5′—>3′ digestion of the transcript. Genes Dev 1994; 8(7):855-66.
31. Wang Z, Kiledjian M. Functional link between the mammalian exosome and mRNA decapping. Cell 2001; 107(6):751-62.
32. Liu H, Rodgers ND, Jiao X et al. The scavenger mRNA decapping enzyme DcpS is a member of the HIT family of pyrophosphatases. EMBO J 2002; 21(17):4699-708.
33. Johnson AW, Kolodner RD. Synthetic lethality of sep1 (xrn1) ski2 and sep1 (xrn1) ski3 mutants of Saccharomyces cerevisiae is independent of killer virus and suggests a general role for these genes in translation control. Mol Cell Biol 1995; 15(5):2719-27.
34. Cao D, Parker R. Computational modeling of eukaryotic mRNA turnover. RNA 2001; 7(9):1192-212.
35. Lebreton A, Tomecki R, Dziembowski A et al. Endonucleolytic RNA cleavage by a eukaryotic exosome. Nature 2008; 456(7224):993-6.
36. Schaeffer D, Tsanova B, Barbas A et al. The exosome contains domains with specific endoribonuclease, exoribonuclease and cytoplasmic mRNA decay activities. Nat Struct Mol Biol 2009; 16(1):56-62.
37. Schneider C, Leung E, Brown J et al. The N-terminal PIN domain of the exosome subunit Rrp44 harbors endonuclease activity and tethers Rrp44 to the yeast core exosome. Nucleic Acids Res 2009; 37(4):1127-40.
38. Dziembowski A, Lorentzen E, Conti E et al. A single subunit, Dis3, is essentially responsible for yeast exosome core activity. Nat Struct Mol Biol 2007; 14(1):15-22.
39. Chen CY, Shyu AB. AU-rich elements: characterization and importance in mRNA degradation. Trends in biochemical sciences 1995; 20(11):465-70.
40. Shaw G, Kamen R. A conserved AU sequence from the 3′ untranslated region of GM-CSF mRNA mediates selective mRNA degradation. Cell 1986; 46(5):659-67.
41. Chen CY, Gherzi R, Ong SE et al. AU binding proteins recruit the exosome to degrade ARE-containing mRNAs. Cell 2001; 107(4):451-64.
42. Gherzi R, Lee KY, Briata P et al. A KH Domain RNA Binding Protein, KSRP, Promotes ARE-Directed mRNA Turnover by Recruiting the Degradation Machinery. Mol Cell 2004; 14(5):571-83.
43. Mukherjee D, Gao M, O'Connor JP et al. The mammalian exosome mediates the efficient degradation of mRNAs that contain AU-rich elements. EMBO J 2002; 21(1-2):165-74.

44. Frischmeyer PA, van Hoof A, O'Donnell K et al. An mRNA surveillance mechanism that eliminates transcripts lacking termination codons. Science 2002; 295(5563):2258-61.
45. Wilson MA, Meaux S, van Hoof A. Diverse aberrancies target yeast mRNAs to cytoplasmic mRNA surveillance pathways. Biochim Biophys Acta. 2008.
46. Shyu AB, Wilkinson MF, van Hoof A. Messenger RNA regulation: to translate or to degrade. EMBO J 2008; 27(3):471-81.
47. Amrani N, Jacobson A. All termination events are not equal: Premature termination in yeast is aberrant and triggers NMD. In: Maquat LE, editor. Nonsense-Mediated mRNA decay. georgetown, TX: Landes Bioscience; 2006. p. 15-26.
48. Maquat LE. NMD in mammalian cells: a history. In: Maquat LE, ed. Nonsense-Mediated mRNA Decay. Georgetown: Landes Bioscience, 2006:43-58.
49. Baker KE, Parker R. Features of nonsense-mediated mRNA decay. In: Maquat LE, ed. Nonsense-Mediated mRNA Decay. Georgetown: Landes Bioscience, 2006:1-14.
50. Leeds P, Peltz SW, Jacobson A et al. The product of the yeast UPF1 gene is required for rapid turnover of mRNAs containing a premature translational termination codon. Genes Dev 1991; 5(12A):2303-14.
51. Leeds P, Wood JM, Lee BS et al. Gene products that promote mRNA turnover in Saccharomyces cerevisiae. Mol Cell Biol 1992; 12(5):2165-77.
52. He F, Jacobson A. Identification of a novel component of the nonsense-mediated mRNA decay pathway by use of an interacting protein screen. Genes Dev 1995; 9(4):437-54.
53. Cui Y, Hagan KW, Zhang S et al. Identification and characterization of genes that are required for the accelerated degradation of mRNAs containing a premature translational termination codon. Genes Dev 1995; 9(4):423-36.
54. Lee BS, Culbertson MR. Identification of an additional gene required for eukaryotic nonsense mRNA turnover. Proc Natl Acad Sci USA 1995; 92(22):10354-8.
55. Muhlrad D, Parker R. Premature translational termination triggers mRNA decapping. Nature 1994; 370(6490):578-81.
56. Takahashi S, Araki Y, Sakuno T et al. Interaction between Ski7p and Upf1p is required for nonsense-mediated 3′-to-5′ mRNA decay in yeast. EMBO J 2003; 22(15):3951-9.
57. Eberle AB, Lykke-Andersen S, Muhlemann O et al. SMG6 promotes endonucleolytic cleavage of nonsense mRNA in human cells. Nat Struct Mol Biol 2009; 16(1):49-55.
58. Huntzinger E, Kashima I, Fauser M et al. SMG6 is the catalytic endonuclease that cleaves mRNAs containing nonsense codons in metazoan. RNA. 2008.
59. Grishok A, Pasquinelli AE, Conte D et al. Genes and mechanisms related to RNA interference regulate expression of the small temporal RNAs that control C. elegans developmental timing. Cell 2001; 106(1):23-34.
60. Hutvagner G, McLachlan J, Pasquinelli AE et al. A cellular function for the RNA-interference enzyme Dicer in the maturation of the let-7 small temporal RNA. Science 2001; 293(5531):834-8.
61. Ketting RF, Fischer SE, Bernstein E et al. Dicer functions in RNA interference and in synthesis of small RNA involved in developmental timing in C. elegans. Genes Dev 2001; 15(20):2654-9.
62. Knight SW, Bass BL. A role for the RNase III enzyme DCR-1 in RNA interference and germ line development in Caenorhabditis elegans. Science 2001; 293(5538):2269-71.
63. Hammond SM, Bernstein E, Beach D et al. An RNA-directed nuclease mediates posttranscriptional gene silencing in Drosophila cells. Nature 2000; 404(6775):293-6.
64. Hammond SM, Boettcher S, Caudy AA et al. Argonaute2, a link between genetic and biochemical analyses of RNAi. Science 2001; 293(5532):1146-50.
65. Martick M, Horan LH, Noller HF et al. A discontinuous hammerhead ribozyme embedded in a mammalian messenger RNA. Nature 2008; 454(7206):899-902.
66. Salehi-Ashtiani K, Luptak A, Litovchick A, Szostak JW. A genomewide search for ribozymes reveals an HDV-like sequence in the human CPEB3 gene. Science 2006; 313(5794):1788-92.
67. Webb CH, Riccitelli NJ, Ruminski DJ, Luptak A. Widespread occurrence of self-cleaving ribozymes. Science 2009; 326(5955):953.
68. Meaux S, van Hoof A. Yeast transcripts cleaved by an internal ribozyme provide new insight into the role of the cap and poly(A) tail in translation and mRNA decay. RNA 2006; 12(7):1323-37.
69. Ball SG, Tirtiaux C, Wickner RB. Genetic Control of L-a and L-(Bc) Dsrna Copy Number in Killer Systems of Saccharomyces cerevisiae. Genetics 1984; 107(2):199-217.
70. Ridley SP, Sommer SS, Wickner RB. Superkiller mutations in Saccharomyces cerevisiae suppress exclusion of M2 double-stranded RNA by L-A-HN and confer cold sensitivity in the presence of M and L-A-HN. Mol Cell Biol 1984; 4(4):761-70.
71. Widner WR, Wickner RB. Evidence that the SKI antiviral system of Saccharomyces cerevisiae acts by blocking expression of viral mRNA. Molecular and cellular biology 1993; 13(7):4331-41.
72. Masison DC, Blanc A, Ribas JC et al. Decoying the cap- mRNA degradation system by a double-stranded RNA virus and poly(A)- mRNA surveillance by a yeast antiviral system. Mol Cell Biol 1995; 15(5):2763-71.
73. Akira S, Uematsu S, Takeuchi O. Pathogen recognition and innate immunity. Cell 2006; 124(4):783-801.

74. Kawai T, Akira S. Innate immune recognition of viral infection. Nat Immunol 2006; 7(2):131-7.
75. Pichlmair A, Reis e Sousa C. Innate recognition of viruses. Immunity 2007; 27(3):370-83.
76. Lu R, Maduro M, Li F et al. Animal virus replication and RNAi-mediated antiviral silencing in Caenorhabditis elegans. Nature 2005; 436(7053):1040-3.
77. Mourrain P, Beclin C, Elmayan T et al. Arabidopsis SGS2 and SGS3 genes are required for posttranscriptional gene silencing and natural virus resistance. Cell 2000; 101(5):533-42.
78. Schott DH, Cureton DK, Whelan SP et al. An antiviral role for the RNA interference machinery in Caenorhabditis elegans. Proc Natl Acad Sci USA 2005; 102(51):18420-4.
79. Wilkins C, Dishongh R, Moore SC et al. RNA interference is an antiviral defence mechanism in Caenorhabditis elegans. Nature 2005; 436(7053):1044-7.
80. Hornung V, Ellegast J, Kim S et al. 5′-Triphosphate RNA is the ligand for RIG-I. Science 2006; 314(5801):994-7.
81. Pichlmair A, Schulz O, Tan CP et al. RIG-I-mediated antiviral responses to single-stranded RNA bearing 5′-phosphates. Science 2006; 314(5801):997-1001.
82. Saito T, Owen DM, Jiang F et al. Innate immunity induced by composition-dependent RIG-I recognition of hepatitis C virus RNA. Nature 2008; 454(7203):523-7.
83. Weir JR, Bonneau F, Hentschel J, Conti E. Structural analysis reveals the characteristic features of Mtr4, a DExH helicase involved in nuclear RNA processing and surveillance. Proc Natl Acad Sci USA 2010; in press.

CHAPTER 8

Rrp6, Rrp47 AND COFACTORS OF THE NUCLEAR EXOSOME

J. Scott Butler*,[1] and Phil Mitchell*,[2]

[1]*Departments of Microbiology and Immunology, and Biochemistry and Biophysics, and Center for RNA Biology: from Genome to Medicine, University of Rochester Medical Center, Rochester, New York, USA;* [2]*Department of Molecular Biology and Biotechnology, University of Sheffield, Sheffield, UK.*
**Correspondence—Emails: scott_butler@urmc.rochester.edu, p.j.mitchell@shef.ac.uk*

Abstract: This chapter reviews the present state of knowledge on the activity of enzymes that function with the RNA exosome in the nucleus. In this compartment, the exosome interacts physically and functionally with the exoribonuclease Rrp6 and several cofactors, most prominently Rrp47 and the TRAMP complex. These interactions decide the fate of RNA precursors from transcription through the formation of mature ribonucleoprotein particles (RNPs) and the export of the RNPs to the cytoplasm. The nuclear exosome catalyzes the formation of the mature 3′ ends of many of these RNAs, but in other cases degrades the RNAs to mononucleotides. Cofactors such as Mpp6, TRAMP and the Nrd1/Nab3 complex play important roles in determining the outcome of the interaction of RNPs with the nuclear exosome. The details that govern the specificity of these decisions remain a rich source for future investigation.

INTRODUCTION

The RNA exosome plays an essential role in the processing and degradation of RNAs in eukaryotic organisms. In the nucleus and the cytoplasm the nine-subunit exosome core, Exo9, and the ribonuclease Dis3/Rrp44 function as a unit, designated Exo10. In *Saccharomyces cerevisiae* this complex interacts physically and functionally with a nucleus specific enzyme, Rrp6, to form the nuclear exosome, Exo11. While the majority of Rrp6 resides in the nucleus in *S. cerevisiae*, evidence suggests its presence in the cytoplasm in humans, *T. brucei* and *A. thaliana*.[1-3] In *S. cerevisiae*, where Rrp6

RNA Exosome, edited by Torben Heick Jensen.

has been studied most extensively, deletion of the sole copy of its gene (*RRP6*) causes a slow growth phenotype at 30°C and extremely poor growth at 37°C.[4] Nevertheless, the fact that deletion of any of the other exosome genes causes lethality has made the use of *rrp6*-Δ strains a valuable tool for the study of exosome defects in nuclear RNA processing. These studies revealed a critical role for Rrp6 in maturation and degradation pathways that include all known classes of RNA. The targeting of Rrp6 and Exo10 to these different RNA processing pathways is specified by interactions with protein co-factors such as Rrp47, Mpp6 and the TRAMP complex. Moreover, studies in *rrp6* mutants, or cells depleted of Rrp6, uncovered the existence of RNA polymerase II transcripts from virtually every part of the genomes of organisms as divergent as yeast, plants and humans. These revelations along with evidence that Rrp6 regulates the levels of specific mRNAs indicate that the nuclear exosome and its co-factors may have key functions in the control of gene expression and organismal development.

STRUCTURE AND ACTIVITY OF Rrp6

Rrp6 belongs to the RNaseD family of the DEDD superfamily of exoribonucleases, which use a two-metal ion mechanism for RNA hydrolysis (Fig. 1).[5] Structure-function studies of Rrp6 proteins with point mutations in the exonuclease domain confirmed the two-metal ion mechanism and suggested that, like the exonuclease domains of DNA polymerases, Rrp6 utilizes a phenylalanine to stabilize the hydroxyl anion intermediate activated for phosphodiester bond cleavage.[6,7] Unlike Dis3/Rrp44, whose activity is attenuated by interaction with Exo9, Rrp6 retains its characteristic properties in the Exo11 complex.[8] Rrp6 contains two HRDC (Helicase RNaseD C-terminal) domains, only one of which was predicted by sequence homology. The Rrp6 HRDC1 domain folds into a characteristic 5-helix structure nearly identical to the homologous portion of *E. coli* RNaseD (Fig. 1).[7,9] Surprisingly, a second HRDC domain appears directly after this in

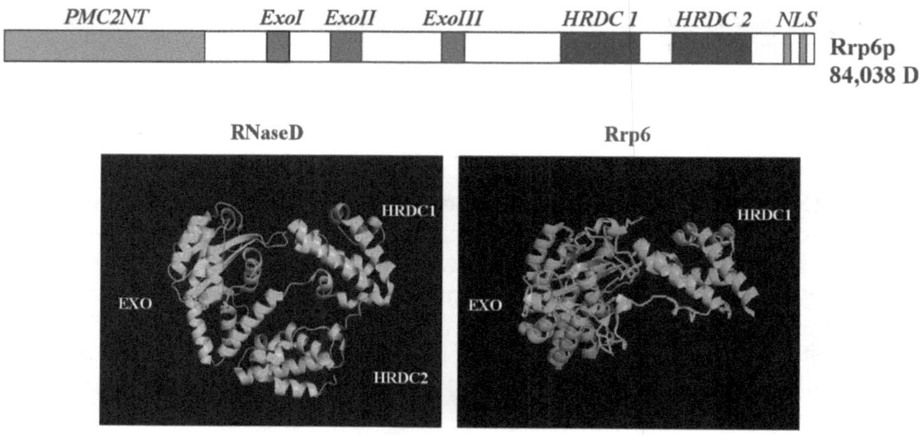

Figure 1. Structure of Rrp6. The top diagram illustrates the polypeptide structure of Rrp6. The bottom panels compare the crystal structures of *E. coli* RNaseD and a N-terminal and C-terminal truncated version of Rrp6 from *S. cerevisiae*. The molecules in the panels were derived from the Protein Data Base using PyMol.

the RNaseD structure, despite a paucity of sequence similarity to HRDC1 or other HRDC domains. Although the polypeptide used for crystal structure analysis of *S. cerevisiae* Rrp6 did not carry HRDC2, it seems reasonable to believe that this region folds to create a similar structure in eukaryotes. Comparison of the activities of *S. cerevisiae* Rrp6 derivatives deleted for either HRDC1 or HRDC2 indicated that HRDC2 plays a critical role in the ability of the protein to interact with Exo10 in vivo.[10] Rrp6 deleted for HRDC2, but not HRDC1, carries out RNA 3′ end processing of pre-snoRNAs and pre-5.8S rRNA (see below), but fails to degrade certain rRNA intermediates that require cooperation between the activities of Rrp6 and Dis3/Rrp44. Thus, the HRDC2 domain appears to facilitate protein-protein interaction between Rrp6 and Exo10. Rrp6 and its eukaryotic homologues carry an N-terminal region (PMC2NT) not found in bacteria. This region is dispensable for interaction with Exo10, but is necessary and sufficient for binding to the Rrp6 cofactor Rrp47/Lrp1 (C1D in humans; see below).[11] Thus, for RNA 3′ end maturation reactions the interaction of Rrp6 with Rrp47 appears more important than the ability of Rrp6 to bind Exo10.

THE ROLE OF Rrp6 IN THE MATURATION OF NONCODING RNAs

Early studies showed that Rrp6 plays a critical role in the maturation of 5.8S pre-RNA.[4] Rrp6 catalyzes the 3′ end trimming of 5.8S pre-rRNA from its 5.8S+30 form to the 6S form, whose end is trimmed to the mature 5.8S rRNA by cytoplasmic Ngl2.[4,12] Rrp6 requires Rrp47/Lrp1 in budding yeast, or its homologue C1D in humans, for efficient trimming of 3′ extended 5.8S precursors.[11,13,14] The 5.8S+30 pre-RNA substrate results from 3′-5′ processing by Exo10 of a longer precursor, 7S pre-rRNA, generated by cleavage at the C2 site within the pre-rRNA internal transcribed spacer.[15] Thus, generation of 6S pre-rRNA in the nucleus requires the concerted 3′-5′ exoribonuclease activities of Dis3/Rrp44 (in the context of Exo10) and Rrp6, suggesting that the Exo11 complex converts 7S to 6S pre-rRNA. However, a mutation that disrupts the interaction between Rrp6 and Exo10 in vivo has no apparent effect on the conversion of 5.8S+30 pre-RNA to 6S pre-rRNA, indicating physical independence of Rrp6 and Exo10 in this step.[10]

Likewise, Rrp6 and Exo10 carry out stepwise 3′ end formation of snoRNAs and snRNAs precursors.[16,17] The independently transcribed snoRNAs and some of the snRNAs are synthesized as 3′ extended precursors that are trimmed in the 3′-5′ direction by Exo10, followed by removal of the last few nucleotides by Rrp6. Like the processing of 7S pre-rRNA to 6S pre-rRNA, the last step, catalyzed by Rrp6, occurs efficiently in an Rrp6 mutant defective for interaction with Exo10, but requires interaction with Rrp47 via the Rrp6 PMC2NT domain.[11,18] Several studies suggest that some portion of mature snoRNAs results from the 3′ trimming of extended, polyadenylated forms of snoRNA precursors, as opposed to their total degradation by an exosome catalyzed surveillance mechanism.[19-21] This raises a longstanding question of how Rrp6 and Exo10 recognize which RNA substrates should be destroyed and which should be trimmed to their normal 3′ ends. Presumably, the answer depends on whether the RNAs have the combined RNA and protein structural information to keep the RNPs in a productive biogenesis pathway.

THE ROLE OF Rrp6 IN mRNA SURVEILLANCE

Rrp6 was discovered by virtue of the fact that loss of function mutations suppress the temperature sensitive growth phenotype of a *S. cerevisiae* strain carrying a mutation (*pap1-1*) in the gene encoding the canonical mRNA poly(A)-polymerase Pap1.[4] Interestingly, loss of Rrp6 activity allowed the accumulation of poly(A)$^+$ mRNAs under conditions where partial inactivation of Pap1 otherwise resulted in the disappearance of most mRNAs.[22] Localization of Rrp6 to the nucleus, demonstration of its hydrolytic exoribonuclease activity and the fact that loss of its activity allowed accumulation of poly(A)+ mRNAs in *pap1-1* strains without changing the rate limiting step in mRNA decay lead to the proposal that the enzyme plays a role in a nuclear mRNA surveillance pathway.[22] Indeed, loss of Rrp6 activity reverses the disappearance of mRNAs caused by defects in other components of the mRNA 3′ end formation pathway, as well as *cis*-acting defects that inhibit polyadenylation.[23-25]

The role of Rrp6 in nuclear mRNA surveillance includes an interesting, but poorly understood function in the accumulation of incompletely processed mRNAs at, or near, their site of transcription.[26] Retention of specific transcripts in single nuclear foci in budding yeast occurs as a result of defects in 3′ end processing, or defects in the formation of export competent RNPs.[27] Strains lacking Rrp6 fail to form these foci and the transcripts exit the nucleus. These observations suggest that, in addition to its role in degrading aberrant transcripts, the presence of Rrp6 slows the transition of pre-mRNAs to export competent RNPs. Indeed, defects in components of the THO/Sub2 complex, which bridges posttranscriptional events with mRNA export, result in mRNA degradation and Rrp6-dependent retention of pre-mRNAs in nuclear foci.[27-29] The mechanistic details of the connection between the THO/Sub2 complex and the nuclear exosome remain unclear, but recent evidence suggests that THO/Sub2 defects inhibit the activity of the mRNA 3′ end processing machinery.[29,30] One interpretation of these findings is that disruption of the transition from pre-mRNA to an export competent RNP may feedback to down regulate mRNA 3′ end processing, thereby exposing transcripts to degradation by Exo11.

Defects in Rrp6 and other exosome components render yeast and human cells hypersensitive to the chemotherapeutic drug 5-fluorouracil (5FU).[31-33] This effect; (i) results from incorporation of the base analogue into RNA, (ii) is abolished by a mutation in Rrp6 that inhibits its degradation, but not by one that inhibits its 3′ end maturation function, (iii) correlates with the accumulation of poly(A)$^+$ RNA degradation intermediates and (iv) is independent of the transcription-coupled DNA repair pathway.[32,34] Moreover, hypersensitivity of *rrp6*-Δ strains to 5FU requires a catalytically active pseudouridine synthetase, Cbf5.[35] Some pseudouridine synthetases cannot convert 5FU into psuedouridine and the incomplete enzymatic reaction results in covalent adducts between the enzyme and 5FU-RNA. These findings suggest that a significant component of 5FU cytotoxicity in *S. cerevisiae* may result from the inability of Rrp6 to degrade RNA molecules trapped in complexes with Cbf5.

THE ROLE OF Rrp6 IN THE REGULATION OF mRNA LEVELS

Rrp6 controls the levels of normal mRNAs as exemplified by the autoregulation of the nuclear poly(A)-binding protein Nab2 activity in budding yeast.[36] The interaction of Nab2 with an oligoadenylate sequence in the 3′ UTR of its own mRNA destabilizes the transcript in a manner dependent on Rrp6 activity. Interestingly, *NAB2* mRNA 3′ end formation occurs by a noncanonical mechanism that requires Exo10 and the Trf4 component of the TRAMP complex.[37] A similar mechanism forms the mature end of the *CTH2* mRNA and, possibly other mRNAs.[38] Rrp6 and the TRAMP complex regulate the levels of histone mRNAs by degrading the transcripts and contributing to their removal at the end of the S-phase of the cell cycle. The requirement for Rrp6 suggests that the negative effect of histone synthesis beyond the end of S-phase requires the rapid destruction of their mRNAs in the nucleus as well as in the cytoplasm.[39,40] In *Drosophila melanogaster*, depletion of Rrp6 leads to mitotic defects that may reflect altered accumulation of mitotic mRNAs.[41] Similarly, loss of Rrp6 activity causes meiotic defects in *Schizosaccharomyces pombe* and evidence suggests that the Mmi protein targets meiosis-specific pre-mRNAs for degradation by Rrp6.[42,43] Rrp6 also participates in a pathway that degrades mRNAs that exit the nucleus slowly; either due to defects in nuclear export, or naturally slow, mRNA-specific export rates.[44] In this pathway, called DRN (Degradation of RNA in the Nucleus), Rrp6 acts in concert with the nuclear RNA cap-binding complex to accelerate the degradation of slowly exported mRNAs.[25] These findings indicate that cells utilize the nuclear mRNA surveillance function of Rrp6 and perhaps Exo11 to regulate the concentration of a number of mRNAs for the purposes of simple feedback control and more complex regulation of cell cycle events. It seems likely that the list of mRNAs whose levels are subject to regulation by Rrp6 and the exosome will continue to grow.

Rrp6 AND THE TRAMP COMPLEX

The surprising discovery of polyadenylated noncoding RNAs and the existence of noncanonical poly(A)-polymerases revealed a role for Rrp6 and Exo10 in a polyadenylation dependent pathway for the degradation of aberrant RNA processing intermediates and transcripts arising from pervasive transcription of the genome. The first clues to the existence of this pathway came from experiments that demonstrated the accumulation of polyadenylated sn- and snoRNAs and pre-rRNAs in budding yeast strains lacking Rrp6.[16,17,45] A key experiment also revealed that the degradation of hypomodified pre-tRNA$_i^{Met}$ requires the activity of Dis3/Rrp44 and Rrp6.[46] Importantly, these studies identified a requirement for the activity of the noncanonical poly(A)-polymerase Trf4 in destruction of the aberrant pre-tRNA$_i^{Met}$. Later studies showed that Trf4 and another noncanonical poly(A)-polymerase, Trf5, exist in complexes with putative RNA-binding proteins Air1 and Air2, as well as with the RNA helicase Mtr4.[47,48] These complexes, called TRAMP4 and TRAMP5 respectively, play an essential role in the degradation of noncoding RNAs in eukaryotes.[49-51] In some cases, these transcripts may arise from pervasive transcription of the genome that produces a surprising array of sense, antisense and intergenic transcripts, in addition to gene-encoding RNAs.[52-54] Many of these RNAs, as well as intermediates in the biogenesis pathways producing sn-/snRNAs and rRNAs, are polyadenylated by the TRAMP complexes, which facilitates their hydrolysis by Rrp6 and Exo10.

This surveillance mechanism ensures the destruction of aberrant RNA processing intermediates and disposes of transcripts that result from pervasive bi-directional transcription initiation by RNA polymerase II.[55,56] While some of the RNAs generated in this manner appear to lack any function, evidence from budding yeast and *Arabidopsis thaliana* suggests that some influence gene regulation, development and gene silencing.[57-60]

Experiments in *S. pombe* revealed a requirement for TRAMP and Rrp6 in the degradation of transcripts arising from silenced heterochromatin and in posttranscriptional control of RNAi-dependent gene silencing.[61-63] TRAMP complexes likely play a critical role in determining the levels of these regulatory transcripts, since polyadenylation by TRAMP enhances RNA degradation by the nuclear exosome. However, how TRAMP distinguishes between RNAs destined for rapid turnover and other stable RNAs remains a major unanswered question.

THE EXOSOME COFACTOR Rrp47

The yeast protein Rrp47 (also known as Lrp1 or yC1D) and its mammalian homologue C1D are eukaryotic, nuclear proteins that bind both RNA and DNA.[11,14,64,65] Research on Rrp47 has focussed on its physical association with the exosome nuclease complex.[13,66,67] and the role of this complex in RNA processing and degradation pathways (for recent reviews of exosome structure and function, see),[52,68-71] while C1D was initially characterised as a DNA-binding protein that functions in transcription and DNA repair.[64,72,73] Nevertheless, there is strong evidence of functional conservation of Rrp47/C1D; both proteins interact directly with the catalytic exosome component Rrp6 (known as PM-Scl100 in humans),[11,14] yeast *rrp47*-Δ mutants show defects in DNA repair[74,75] and similar defects on 5.8S rRNA maturation are observed in yeast *rrp47*-Δ and *rrp6*-Δ mutants and upon depletion of C1D in mammalian cells.[4,13,14]

STRUCTURE OF Rrp47

Rrp47 and C1D are small, basic proteins of 21kDa and 16kDa, respectively. There is no detailed structural data currently available for these proteins but sequence homology and secondary structure prediction programmes[76] suggest that the first ~120 amino acid residues of Rrp47 comprise a conserved N-terminal α-helical domain, while the C-terminal region is less structured and more variable. The C-terminus of Rrp47 and homologous proteins is rich in basic residues and presumably contributes to RNA binding.[11,77] Rrp47 and C1D can both be phosphorylated in vitro[73,78] but the functional relevance of this is not clear. In contrast to other exosome proteins including Rrp6, Rrp47 has not been observed as a phosphoprotein in mass spectrometric analyses of exosome complexes[79,80] or in orthophosphate in vivo labeling experiments (our unpublished data).

The conserved N-terminal region of Rrp47/C1D spans the bioinformatically defined Sas10/C1D domain (residues 10-89 of Rrp47)[81] also present in Sas10 (also known as Utp3) and Lcp5, two U3 small nucleolar RNA (snoRNA)-associated proteins that are components of the small subunit processosome or pre-90S subunit.[82-86] Overexpression of Sas10 partially suppressed silencing of transcriptionally repressed chromatin in a Sir-independent manner,[87] an effect that has also been reported for mutants in the exosome and TRAMP complexes.[88,89] One possibility is that overexpression of Sas10 may (partially) titrate out the complex between Rrp47 and Rrp6. A Sas10/C1D domain is also found in the protein neuroguidin, an eIF4E-binding protein that is required for cytoplasmic polyadenylation-dependent translational control in neuronal cells.[90] Sas10, Lcp5 and neuroguidin also have basic regions at their C-termini that may contribute to

RNA binding but whether the Sas10/C1D domain represents an RNA-binding domain requires further experimentation.

BIOCHEMICAL ACTIVITIES OF Rrp47

Rrp47 and C1D are associated with Rrp6-containing exosome complexes[13,67] and interact directly with Rrp6 in vitro.[11,14] Binding occurs via the N-terminal PMC2NT domain of Rrp6[91] and the N-terminal α-helical region of Rrp47 (our unpublished observations). Westerns of epitope-tagged proteins show that Rrp47 and Rrp6 are expressed at comparable levels in yeast,[11] Rrp6 being present at approximately 2,000 molecules per cell.[92] Normal expression levels of Rrp47 in yeast are dependent upon this interaction with Rrp6.[11] Therefore, the interpretation of data from experiments using *rrp6-Δ* null alleles should take into account that Rrp47 expression levels are also significantly affected.

Rrp47 and C1D show specificity for structured RNA in vitro.[11,14] Consistent with an important role for Rrp47 RNA binding activity in vivo, RNAs that accumulate in *rrp47-Δ* mutants, such as the 3′ extended 5.8S rRNA precursor, are predicted to have double stranded regions at their 3′ termini.[93] The dissociation constant of yeast Rrp47 for RNA and DNA is approximately 1 μM (calculating the protein concentration based on the predicted molecular weight of the monomeric protein).[11] C1D is reported to have "exceptional DNA affinity"[64] but the dissociation constant for DNA or RNA has not been reported and it is not clear whether the native protein has a preference for DNA or RNA substrates. Assuming there are approximately 2,000 molecules of Rrp47 per yeast cell and the protein is distributed evenly throughout the nucleus, which has a volume of approximately 3 μm,[94] the intracellular concentration of Rrp47 is close to its dissociation constant. Therefore, small regulatory changes in the effective concentration of Rrp47 would impact strongly on the efficiency with which this protein would bind its target substrates. Notably, overexpression of C1D is toxic.[95]

THE ROLE OF Rrp47 IN RNA PROCESSING AND DEGRADATION

Yeast strains lacking Rrp47 or Rrp6 accumulate a common set of cellular RNAs that, in wild type cells, are normally efficiently processed to mature RNAs or rapidly degraded. Rrp47 is not required for normal Rrp6 expression levels or the association of Rrp6 with exosome complexes.[13] Since a trimeric complex can be formed between Rrp47, Rrp6 and RNA, it has been suggested that Rrp47 promotes substrate binding to Rrp6.[11] Purified Rrp6 degrades unstructured RNA efficiently but is blocked by stem loop structures,[8,22] suggesting that Rrp47 may facilitate degradation of structured RNA. Rrp47/C1D does not have any sequence homology to RNA helicases. Rrp47 might simply increase the retention time of Rrp6 on structured substrates or it may function sterically by positioning the 3′ end of structured RNA close to the catalytic centre of Rrp6, as has been proposed for the C'-terminal HRDC domain of RNase D.[9] Notably, while the pattern of stable RNA precursors observed in *rrp47-Δ* and *rrp6-Δ* strains by northern blot analyses are largely indistinguishable,[13] distinct effects are seen in these mutants upon analysis of global mRNA profiles.[75,96] This suggests that Rrp6 can function independently of Rrp47 on some substrates.

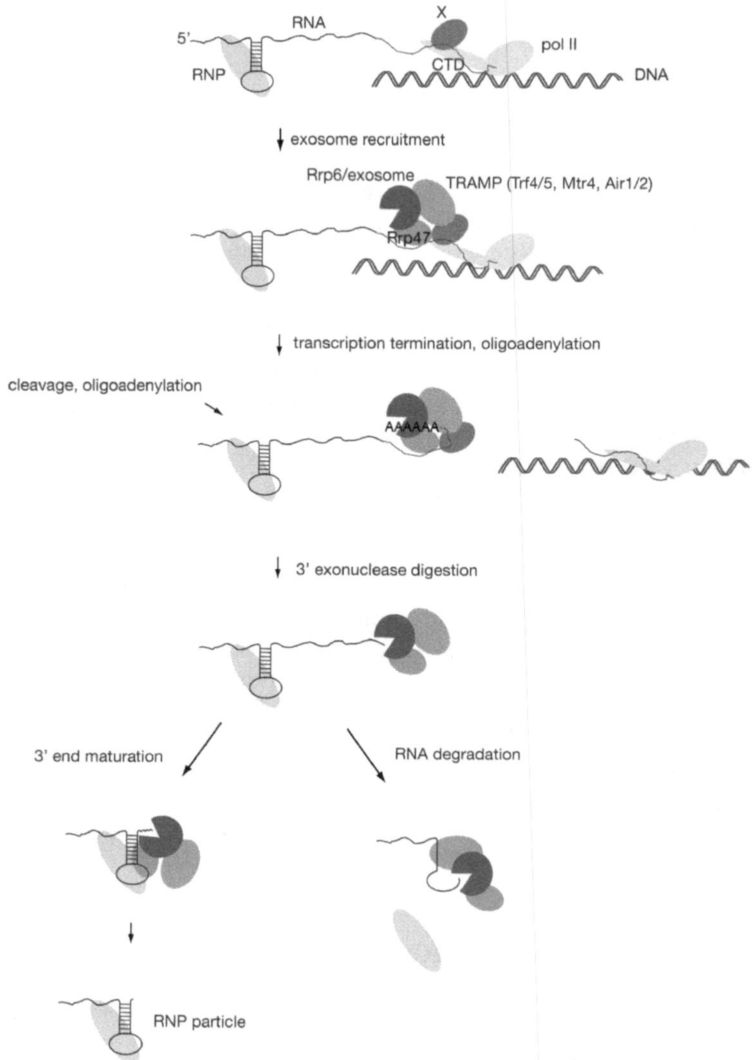

Figure 2. Substrate recognition and degradation by the nuclear exosome. The schematic depicts the general pathway of exosome recruitment and substrate digestion/processing by the nuclear exosome complex. The exosome is recruited to its RNA substrates (either polymerase-engaged transcription termination complexes or RNP particles with accessible 3′ ends) through a poorly understood set of interactions that, in most cases, will involve both RNA binding with undetermined sequences or structures and protein recognition with poorly characterised partner proteins (labelled "X"). The DNA binding activity of Rrp47 may also promote recruitment of the exosome to termination regions. Degradation/processing of the RNA can be promoted by Trf4- or Trf5-mediated oligoadenylation and by the RNA helicase activity of Mtr4, all of which are components of the TRAMP complex, while Rrp47, Mpp6 and the Air proteins contribute to RNA binding. RNA hydrolysis by the Rrp6 and Rrp44 exonucleases progresses until enzyme activity is blocked by the preassembled RNP particle, allowing 3′ end maturation. In the absence of correct RNP particle assembly, the RNA is degraded completely by the exosome and TRAMP complexes. Other substrates are generated by endonucleolytic cleavage or arise through delayed transcription/processing events. For some substrates, Rrp6 can function together with Rrp47 and TRAMP, independently of the core exosome complex.

Rrp47 might also function in substrate recognition, either through interaction with RNA or other processing factors (Fig. 2). Interactions between the Nrd1 termination complex and the exosome and TRAMP complexes are thought to recruit the degradation/processing machinery to some of its RNA substrates[97] and Rrp47 is present in such large RNA processing/degradation complexes.[98] Nrd1 and the associated protein Nab3 are RNA-binding proteins that bind cooperatively to multiple copies of tetrameric recognition sequences present in the terminator regions of target transcripts.[99] The *rrp47*-Δ mutation is synthetic lethal with the RNA-binding defective *nrd1-102* mutation,[100] suggesting that the RNA-binding activity of Rrp47 may play an important role in the recruitment of the exosome to the termination complex. It has also been suggested that the interaction between Rrp47 and DNA may help position exosome complexes at transcriptionally active sites.[101]

Rrp47/C1D is well annotated as a DNA repair factor. Yeast strains lacking Rrp47 or Rrp6 are defective in the degradation of specific mRNA transcripts upon exposure to UV irradiation and in the repair of UV irradiation-induced cyclobutane dimers and *rrp6*-Δ mutants are synergistically sick in combination with *rad26*-Δ mutants.[75] The sensitivity of *rrp6*-Δ and *rrp47*-Δ mutants to UV irradiation may reflect a quality control function of the exosome to eliminate aberrant transcripts that arise as a result of DNA damage in transcription-coupled DNA repair.[102]

THE EXOSOME COFACTOR MPP6

Mpp6 (MPP6 in humans) shares a number of similarities with Rrp47, both being exosome-associated RNA-binding proteins that are small and basic in nature. The proteins have distinct substrate specificities, however, Rrp47 recognising structured RNA while Mpp6 binds pyrimidine-rich sequences.[11,14,98,103] While Rrp47 is functionally linked to the activity of Rrp6, the role of Mpp6 in exosome function is less clear.

MPP6 was first identified as a nuclear protein that is phosphorylated and distributed throughout the cell during mitosis.[104] MPP6 copurifies with exosome complexes that contain the Rrp6 homologue PM-Scl100 and hMtr4 but which lack the Rrp44/Dis3 homologue.[105] Thus, the interaction between MPP6 and the exosome is independent of Rrp44. Interactions between MPP6 and PM-Scl100 or hMtr4 were first suggested by two-hybrid interaction[106] and subsequently shown to be direct by pull-down experiments using recombinant and in vitro translated proteins.[14] The binding sites within Rrp6 for MPP6 and C1D apparently do not overlap, since a stable trimeric complex could be assembled.

Depletion of MPP6 caused an accumulation of 3′ extended forms of 5.8S rRNA similar to that seen upon loss of exosome function.[103] Notably, the 3′ extended 5.8S pre-rRNA that accumulates upon depletion of MPP6 contains a pyrimidine-rich sequence at its 3′ end, suggesting that MPP6 might target the exosome to this substrate. Consistent with a role in pre-rRNA processing, both the mammalian and yeast proteins are found associated with large ribosome-containing complexes.[103,107]

The yeast protein Mpp6 was first demonstrated to be associated with the exosome complex in a global proteomics study[86] and subsequently shown to be required for viability of *rrp47*-Δ mutants.[98] In common with other exosome mutants, *mpp6*-Δ strains showed defects in 5.8S rRNA maturation, accumulated the IGS1-R rDNA intergenic transcript and suppressed the loss of mRNA observed in *rna14-1* and *prp2-1* strains.[23,88,89,98,108] Notably, the accumulation of the *NEL025C* CUT upon loss of both Mpp6 and Rrp47 was significantly greater than each single mutant,[98] suggesting that an exacerbated accumulation of regulatory RNA transcripts might contribute to the synthetic lethality of *rrp47*-Δ *mpp6*-Δ strains.

The *mpp6*-Δ mutation is also synthetic lethal with the *rrp6*-Δ mutation.[98] This implies that Mpp6 and Rrp47 do not function in a functionally redundant manner to target substrates to Rrp6. It has been proposed that Mpp6 might promote the activity of the other exosome catalytic subunit, Rrp44.[98] Given that the human MPP6 protein physically interacts with PM-Scl100 and hMtr4,[14] that it is associated with the exosome in the absence of Rrp44[105] and that the yeast *mpp6*-Δ mutation shows strong genetic interactions with *air1*-Δ mutations,[98,109] another possibility is that Mpp6 promotes the functional coupling between Rrp6 and the TRAMP complex.

CONCLUSION

The recent discovery of diverse noncoding RNAs that are stabilised in the absence of Rrp6[55,56,110] has dramatically increased the number of substrates known to be processed or degraded by the nuclear exosome. The molecular mechanisms by which the exosome is targeted to its substrate molecules and how RNA processing substrates are distinguished from those targeted to complete degradation, however, remain largely unresolved. These outstanding questions address fundamental issues concerning the links between transcription termination, RNA processing and RNP particle assembly and the distinction between RNA processing events and RNA surveillance. A general point appears to be the redundancy seen for nucleolytic, polymerase and RNA-binding activities in exosome-mediated RNA 3′ processing that presumably facilitate efficient processing and provide a system that is well buffered against genetic modulation.

ACKNOWLEDGEMENTS

S.B. was supported by a grant from the National Science Foundation (MCB-0817324) and P.M. was supported by a research grant from the Wellcome Trust (088306/Z/09/Z).

REFERENCES

1. Haile S, Estevez A, Clayton C. A role for the exosome in the in vivo degradation of unstable mRNAs. RNA 2003; 9:1491-1501.
2. Lange H, Holec S, Cognat V et al. Degradation of a polyadenylated rRNA maturation by-product involves one of the three RRP6-like proteins in Arabidopsis thaliana. Mol Cell Biol 2008; 28:3038-3044.
3. Lejeune F, Li X, Maquat L. Nonsense-mediated mRNA decay in mammalian cells involves decapping, deadenylating and exonucleolytic activities. Mol Cell 2003; 12:675-687.
4. Briggs M, Burkard K, Butler J. Rrp6p, the yeast homologue of the human PM-Scl 100-kDa autoantigen, is essential for efficient 5.8 S rRNA 3′ end formation. J Biol Chem 1998; 273:13255-13263.
5. Zuo Y, Deutscher M. Exoribonuclease superfamilies: structural analysis and phylogenetic distribution. Nucleic Acids Res 2001; 29:1017-1026.
6. Phillips S, Butler J. Contribution of domain structure to the RNA 3′ end processing and degradation functions of the nuclear exosome subunit Rrp6p. RNA 2003; 9:1098-1107.

7. Midtgaard S, Assenholt J, Jonstrup A et al. Structure of the nuclear exosome component Rrp6p reveals an interplay between the active site and the HRDC domain. Proc Natl Acad Sci USA 2006; 103:11898-11903.
8. Liu Q, Greimann J, Lima C. Reconstitution, activities and structure of the eukaryotic RNA exosome. Cell 2006; 127:1223-1237.
9. Zuo Y, Wang Y, Malhotra A. Crystal structure of Escherichia coli RNase D, an exoribonuclease involved in structured RNA processing. Structure 2005; 13:973-984.
10. Callahan K, Butler J. Evidence for core exosome independent function of the nuclear exoribonuclease Rrp6p. Nucleic Acids Res 2008; 36:6645-6655.
11. Stead J, Costello J, Livingstone M et al. The PMC2NT domain of the catalytic exosome subunit Rrp6p provides the interface for binding with its cofactor Rrp47p, a nucleic acid-binding protein. Nucleic Acids Res 2007; 35:5556-5567.
12. Thomson E, Tollervey D. The final step in 5.8S rRNA processing is cytoplasmic in Saccharomyces cerevisiae. Mol Cell Biol 2010; 30:976-984.
13. Mitchell P, Petfalski E, Houalla R et al. Rrp47p is an exosome-associated protein required for the 3′ processing of stable RNAs. Mol Cell Biol 2003; 23:6982-6992.
14. Schilders G, van Dijk E, Pruijn G. C1D and hMtr4p associate with the human exosome subunit PM/Scl-100 and are involved in prerRNA processing. Nucleic Acids Res 2007; 35:2564-2572.
15. Allmang C, Petfalski E, Podtelejnikov A et al. The yeast exosome and human PM-Scl are related complexes of 3′->5′ exonucleases. Genes Dev 1999; 13:2148-2158.
16. van Hoof A, Lennertz P, Parker R. Yeast exosome mutants accumulate 3′-extended polyadenylated forms of U4 small nuclear RNA and small nucleolar RNAs. Mol Cell Biol 2000; 20:441-452.
17. Allmang C, Kufel J, Chanfreau G et al. Functions of the exosome in rRNA, snoRNA and snRNA synthesis. EMBO J 1999; 18:5399-5410.
18. Callahan K, Butler J. The TRAMP complex enhances RNA degradation by the nuclear exosome component Rrp6. J Biol Chem 2009; 285:3540-3547.
19. Egecioglu D, Henras A, Chanfreau G. Contributions of Trf4p- and Trf5p-dependent polyadenylation to the processing and degradative functions of the yeast nuclear exosome. RNA 2006; 12:26-32.
20. Grzechnik P, Kufel J. Polyadenylation linked to transcription termination directs the processing of snoRNA precursors in yeast. Mol Cell 2008; 32:247-258.
21. Lemay J, D'Amours A, Lemieux C et al. The nuclear poly(A)-binding protein interacts with the exosome to promote synthesis of noncoding small nucleolar RNAs. Mol Cell 2010; 37:34-45.
22. Burkard K, Butler J. A nuclear 3′-5′ exonuclease involved in mRNA degradation interacts with Poly(A) polymerase and the hnRNA protein Npl3p. Mol Cell Biol 2000; 20:604-616.
23. Torchet C, Bousquet-Antonelli C, Milligan L et al. Processing of 3′-extended read-through transcripts by the exosome can generate functional mRNAs. Mol Cell 2002; 9:1285-1296.
24. Milligan L, Torchet C, Allmang C et al. A nuclear surveillance pathway for mRNAs with defective polyadenylation. Mol Cell Biol 2005; 25:9996-10004.
25. Das B, Butler J, Sherman F. Degradation of normal mRNA in the nucleus of Saccharomyces cerevisiae. Mol Cell Biol 2003; 23:5502-5515.
26. Hilleren P, McCarthy T, Rosbash M et al. Quality control of mRNA 3′-end processing is linked to the nuclear exosome. Nature 2001; 413:538-542.
27. Libri D, Dower K, Boulay J et al. Interactions between mRNA export commitment, 3′-end quality control and nuclear degradation. Mol Cell Biol 2002; 22:8254-8266.
28. Rougemaille M, Gudipati R, Olesen J et al. Dissecting mechanisms of nuclear mRNA surveillance in THO/sub2 complex mutants. EMBO J 2007; 26:2317-2326.
29. Saguez C, Schmid M, Olesen J et al. Nuclear mRNA surveillance in THO/sub2 mutants is triggered by inefficient polyadenylation. Mol Cell 2008; 31:91-103.
30. Qu X, Lykke-Andersen S, Nasser T et al. Assembly of an export-competent mRNP is needed for efficient release of the 3′-end processing complex after polyadenylation. Mol Cell Biol 2009; 29:5327-5338.
31. Lum P, Armour C, Stepaniants S et al. Discovering modes of action for therapeutic compounds using a genome-wide screen of yeast heterozygotes. Cell 2004; 116:121-137.
32. Fang F, Hoskins J, Butler J. 5-fluorouracil enhances exosome-dependent accumulation of polyadenylated rRNAs. Mol Cell Biol 2004; 24:10766-10776.
33. Kammler S, Lykke-Andersen S, Jensen T. The RNA exosome component hRrp6 is a target for 5-fluorouracil in human cells. Mol Cancer Res 2008; 6:990-995.
34. Hoskins J, Scott Butler J. Evidence for distinct DNA- and RNA-based mechanisms of 5-fluorouracil cytotoxicity in Saccharomyces cerevisiae. Yeast 2007; 24:861-870.
35. Hoskins J, Butler J. RNA-based 5-fluorouracil toxicity requires the pseudouridylation activity of Cbf5p. Genetics 2008; 179:323-330.
36. Roth K, Wolf M, Rossi M et al. The nuclear exosome contributes to autogenous control of NAB2 mRNA levels. Mol Cell Biol 2005; 25:1577-1585.

37. Roth K, Byam J, Fang F et al. Regulation of NAB2 mRNA 3′-end formation requires the core exosome and the Trf4p component of the TRAMP complex. RNA 2009; 15:1045-1058.
38. Ciais D, Bohnsack M, Tollervey D. The mRNA encoding the yeast ARE-binding protein Cth2 is generated by a novel 3′ processing pathway. Nucleic Acids Res 2008; 36:3075-3084.
39. Reis C, Campbell J. Contribution of Trf4/5 and the nuclear exosome to genome stability through regulation of histone mRNA levels in Saccharomyces cerevisiae. Genetics 2007; 175:993-1010.
40. Canavan R, Bond U. Deletion of the nuclear exosome component RRP6 leads to continued accumulation of the histone mRNA HTB1 in S-phase of the cell cycle in Saccharomyces cerevisiae. Nucleic Acids Res 2007; 35:6268-6279.
41. Graham A, Kiss D, Andrulis E. Core exosome-independent roles for Rrp6 in cell cycle progression. Mol Biol Cell 2009; 20:2242-2253.
42. McPheeters D, Cremona N, Sunder S et al. A complex gene regulatory mechanism that operates at the nexus of multiple RNA processing decisions. Nat Struct Mol Biol 2009; 16:255-264.
43. Harigaya Y, Tanaka H, Yamanaka S et al. Selective elimination of messenger RNA prevents an incidence of untimely meiosis. Nature 2006; 442:45-50.
44. Kuai L, Das B, Sherman F. A nuclear degradation pathway controls the abundance of normal mRNAs in Saccharomyces cerevisiae. Proc Natl Acad Sci USA 2005; 102:13962-13967.
45. Kuai L, Fang F, Butler J et al. Polyadenylation of rRNA in Saccharomyces cerevisiae. Proc Natl Acad Sci USA 2004; 101:8581-8586.
46. Kadaba S, Krueger A, Trice T et al. Nuclear surveillance and degradation of hypomodified initiator tRNAMet in S. cerevisiae. Genes Dev 2004; 18:1227-1240.
47. Vanácová S, Wolf J, Martin G et al. A new yeast poly(A) polymerase complex involved in RNA quality control. PLoS Biol 2005; 3:e189.
48. LaCava J, Houseley J, Saveanu C et al. RNA degradation by the exosome is promoted by a nuclear polyadenylation complex. Cell 2005; 121:713-724.
49. Houseley J, Tollervey D. Yeast Trf5p is a nuclear poly(A) polymerase. EMBO Rep 2006; 7:205-211.
50. Win T, Draper S, Read R et al. Requirement of fission yeast Cid14 in polyadenylation of rRNAs. Mol Cell Biol 2006; 26:1710-1721.
51. Wyers F, Rougemaille M, Badis G et al. Cryptic pol II transcripts are degraded by a nuclear quality control pathway involving a new poly(A) polymerase. Cell 2005; 121:725-737.
52. Belostotsky D. Exosome complex and pervasive transcription in eukaryotic genomes. Curr Opin Cell Biol 2009; 21:352-358.
53. Jacquier A. The complex eukaryotic transcriptome: unexpected pervasive transcription and novel small RNAs. Nat Rev Genet 2009; 10:833-844.
54. Kapranov P, Willingham A, Gingeras T. Genome-wide transcription and the implications for genomic organization. Nat Rev Genet 2007; 8:413-423.
55. Neil H, Malabat C, d'Aubenton-Carafa Y et al. Widespread bidirectional promoters are the major source of cryptic transcripts in yeast. Nature 2009; 457:1038-1042.
56. Xu Z, Wei W, Gagneur J et al. Bidirectional promoters generate pervasive transcription in yeast. Nature 2009; 457:1033-1037.
57. Uhler J, Hertel C, Svejstrup J. A role for noncoding transcription in activation of the yeast PHO5 gene. Proc Natl Acad Sci USA 2007; 104:8011-8016.
58. Houseley J, Rubbi L, Grunstein M et al. A ncRNA modulates histone modification and mRNA induction in the yeast GAL gene cluster. Mol Cell 2008; 32:685-695.
59. Camblong J, Iglesias N, Fickentscher C et al. Antisense RNA stabilization induces transcriptional gene silencing via histone deacetylation in S. cerevisiae. Cell 2007; 131:706-717.
60. Chekanova J, Gregory B, Reverdatto S et al. Genome-wide high-resolution mapping of exosome substrates reveals hidden features in the Arabidopsis transcriptome. Cell 2007; 131:1340-1353.
61. Bühler M, Spies N, Bartel D et al. TRAMP-mediated RNA surveillance prevents spurious entry of RNAs into the Schizosaccharomyces pombe siRNA pathway. Nat Struct Mol Biol 2008; 15:1015-1023.
62. Wang S, Stevenson A, Kearsey S et al. Global role for polyadenylation-assisted nuclear RNA degradation in posttranscriptional gene silencing. Mol Cell Biol 2008; 28:656-665.
63. Bühler M, Haas W, Gygi S et al. RNAi-dependent and -independent RNA turnover mechanisms contribute to heterochromatic gene silencing. Cell 2007; 129:707-721.
64. Nehls P, Keck T, Greferath R et al cDNA cloning, recombinant expression and characterization of polypetides with exceptional DNA affinity. Nucleic Acids Res 1998; 26:1160-1166.
65. Kumar A, Agarwal S, Heyman J et al. Subcellular localization of the yeast proteome. Genes Dev 2002; 16:707-719.
66. Gavin A, Bösche M, Krause R et al. Functional organization of the yeast proteome by systematic analysis of protein complexes. Nature 2002; 415:141-147.

67. Peng W, Robinson M, Mnaimneh S et al. A panoramic view of yeast noncoding RNA processing. Cell 2003; 113:919-933.
68. Vanacova S, Stefl R. The exosome and RNA quality control in the nucleus. EMBO Rep 2007; 8:651-657.
69. Lebreton A, Séraphin B. Exosome-mediated quality control: substrate recruitment and molecular activity. Biochim Biophys Acta 2008; 1779:558-565.
70. Schmid M, Jensen T. The exosome: a multipurpose RNA-decay machine. Trends Biochem Sci 2008; 33:501-510.
71. Lykke-Andersen S, Brodersen D, Jensen T. Origins and activities of the eukaryotic exosome. J Cell Sci 2009; 122:1487-1494.
72. Zamir I, Dawson J, Lavinsky R et al. Cloning and characterization of a corepressor and potential component of the nuclear hormone receptor repression complex. Proc Natl Acad Sci USA 1997; 94:14400-14405.
73. Yavuzer U, Smith G, Bliss T et al. DNA end-independent activation of DNA-PK mediated via association with the DNA-binding protein C1D. Genes Dev 1998; 12:2188-2199.
74. Erdemir T, Bilican B, Cagatay T et al. Saccharomyces cerevisiae C1D is implicated in both nonhomologous DNA end joining and homologous recombination. Mol Microbiol 2002; 46:947-957.
75. Hieronymus H, Yu M, Silver P. Genome-wide mRNA surveillance is coupled to mRNA export. Genes Dev 2004; 18:2652-2662.
76. Kelley L, Sternberg M. Protein structure prediction on the Web: a case study using the Phyre server. Nat Protoc 2009; 4:363-371.
77. Spriggs R, Murakami Y, Nakamura H et al. Protein function annotation from sequence: prediction of residues interacting with RNA. Bioinformatics 2009; 25:1492-1497.
78. Ptacek J, Devgan G, Michaud G et al. Global analysis of protein phosphorylation in yeast. Nature 2005; 438:679-684.
79. Synowsky S, van den Heuvel R, Mohammed S et al. Probing genuine strong interactions and posttranslational modifications in the heterogeneous yeast exosome protein complex. Mol Cell Proteomics 2006; 5:1581-1592.
80. Synowsky S, van Wijk M, Raijmakers R et al. Comparative Multiplexed Mass Spectrometric Analyses of Endogenously Expressed Yeast Nuclear and Cytoplasmic Exosomes. J Mol Biol 2008; 385:1300-1313.
81. Finn R, Mistry J, Tate J et al. The Pfam protein families database. Nucleic Acids Res 2010; 38:D211-D222.
82. Wiederkehr T, Prétôt R, Minvielle-Sebastia L. Synthetic lethal interactions with conditional poly(A) polymerase alleles identify LCP5, a gene involved in 18S rRNA maturation. RNA 1998; 4:1357-1372.
83. Uetz P, Giot L, Cagney G et al. A comprehensive analysis of protein-protein interactions in Saccharomyces cerevisiae. Nature 2000; 403:623-627.
84. Dragon F, Gallagher J, Compagnone-Post P et al. A large nucleolar U3 ribonucleoprotein required for 18S ribosomal RNA biogenesis. Nature 2002; 417:967-970.
85. Hoang T, Peng W, Vanrobays E et al. Esf2p, a U3-associated factor required for small-subunit processome assembly and compaction. Mol Cell Biol 2005; 25:5523-5534.
86. Krogan N, Cagney G, Yu H et al. Global landscape of protein complexes in the yeast Saccharomyces cerevisiae. Nature 2006; 440:637-643.
87. Kamakaka R, Rine J. Sir- and silencer-independent disruption of silencing in Saccharomyces by Sas10p. Genetics 1998; 149:903-914.
88. Vasiljeva L, Kim M, Terzi N et al. Transcription termination and RNA degradation contribute to silencing of RNA polymerase II transcription within heterochromatin. Mol Cell 2008; 29:313-323.
89. Houseley J, Kotovic K, El Hage A et al. Trf4 targets ncRNAs from telomeric and rDNA spacer regions and functions in rDNA copy number control. EMBO J 2007; 26:4996-5006.
90. Jung M, Lorenz L, Richter J. Translational control by neuroguidin, a eukaryotic initiation factor 4E and CPEB binding protein. Mol Cell Biol 2006; 26:4277-4287.
91. Staub E, Fiziev P, Rosenthal A et al. Insights into the evolution of the nucleolus by an analysis of its protein domain repertoire. Bioessays 2004; 26:567-581.
92. Ghaemmaghami S, Huh W, Bower K et al. Global analysis of protein expression in yeast. Nature 2003; 425:737-741.
93. Yeh L, Lee J. Structural analysis of the internal transcribed spacer 2 of the precursor ribosomal RNA from Saccharomyces cerevisiae. J Mol Biol 1990; 211:699-712.
94. Jorgensen P, Edgington N, Schneider B et al. The size of the nucleus increases as yeast cells grow. Mol Biol Cell 2007; 18:3523-3532.
95. Rothbarth K, Spiess E, Juodka B et al. Induction of apoptosis by overexpression of the DNA-binding and DNA-PK-activating protein C1D. J Cell Sci 1999; 112:2223-2232.
96. Houalla R, Devaux F, Fatica A et al. Microarray detection of novel nuclear RNA substrates for the exosome. Yeast 2006; 23:439-454.

97. Vasiljeva L, Buratowski S. Nrd1 interacts with the nuclear exosome for 3′ processing of RNA polymerase II transcripts. Mol Cell 2006; 21:239-248.
98. Milligan L, Decourty L, Saveanu C et al. A yeast exosome cofactor, Mpp6, functions in RNA surveillance and in the degradation of noncoding RNA transcripts. Mol Cell Biol 2008; 28:5446-5457.
99. Carroll K, Ghirlando R, Ames J et al. Interaction of yeast RNA-binding proteins Nrd1 and Nab3 with RNA polymerase II terminator elements. RNA 2007; 13:361-373.
100. Arigo J, Eyler D, Carroll K et al. Termination of cryptic unstable transcripts is directed by yeast RNA-binding proteins Nrd1 and Nab3. Mol Cell 2006; 23:841-851.
101. Houseley J, Tollervey D. The nuclear RNA surveillance machinery: The link between ncRNAs and genome structure in budding yeast? Biochim Biophys Acta 2008; 1779:239-246.
102. Mirkin N, Fonseca D, Mohammed S et al. The 3′ processing factor CstF functions in the DNA repair response. Nucleic Acids Res 2008; 36:1792-1804.
103. Schilders G, Raijmakers R, Raats J et al. MPP6 is an exosome-associated RNA-binding protein involved in 5.8S rRNA maturation. Nucleic Acids Res 2005; 33:6795-6804.
104. Matsumoto-Taniura N, Pirollet F, Monroe R et al. Identification of novel M phase phosphoproteins by expression cloning. Mol Biol Cell 1996; 7:1455-1469.
105. Chen C, Gherzi R, Ong S et al. AU binding proteins recruit the exosome to degrade ARE-containing mRNAs. Cell 2001; 107:451-464.
106. Lehner B, Sanderson C. A protein interaction framework for human mRNA degradation. Genome Res 2004; 14:1315-1323.
107. Fleischer T, Weaver C, McAfee K et al. Systematic identification and functional screens of uncharacterized proteins associated with eukaryotic ribosomal complexes. Genes Dev 2006; 20:1294-1307.
108. Bousquet-Antonelli C, Presutti C, Tollervey D. Identification of a regulated pathway for nuclear premRNA turnover. Cell 2000; 102:765-775.
109. Wilmes G, Bergkessel M, Bandyopadhyay S et al. A genetic interaction map of RNA-processing factors reveals links between Sem1/Dss1-containing complexes and mRNA export and splicing. Mol Cell 2008; 32:735-746.
110. Preker P, Nielsen J, Kammler S et al. RNA exosome depletion reveals transcription upstream of active human promoters. Science 2008; 322:1851-1854.

CHAPTER 9

THE EXOSOME AND HETEROCHROMATIN
Multilevel Regulation of Gene Silencing

Sarah Coy and Lidia Vasiljeva*
Department of Biochemistry, University of Oxford, South Parks Road, Oxford, UK.
**Corresponding Author: Lidia Vasilieva—Email: lidia.vasilieva@bioch.ox.ac.uk*

Abstract: Heterochromatic silencing is important for repressing gene expression, protecting cells against viral invasion, maintaining DNA integrity and for proper chromosome segregation. Recently, it has become apparent that expression of eukaryotic genomes is far more complex than had previously been anticipated. Strikingly, it has emerged that most of the genome is transcribed including intergenic regions and heterochromatin, calling for us to re-address the question of how gene silencing is regulated and re-evaluate the concept of heterochromatic regions of the genome being transcriptionally inactive. Although heterochromatic silencing can be regulated at the transcriptional level, RNA degrading activities supplied either by the exosome complex or RNAi also significantly contribute to this process. The exosome also regulates noncoding RNAs (ncRNAs) involved in the establishment of heterochromatin, further underscoring its role as the major cellular machinery involved in RNA processing and turn-over. This multilevel control of the transcriptome may be utilized to ensure greater accuracy of gene expression and allow distinction between functional transcripts and background noise. In this chapter, we will discuss the regulation of gene silencing across species, with special emphasis on the exosome's contribution to the process. We will also discuss the links between transcriptional and posttranscriptional mechanisms for gene silencing and their impact on the regulation of eukaryotic transcriptomes.

INTRODUCTION

Recently, genome-wide studies have revealed that more than 90% of the genome is transcribed in eukaryotes.[1-3] These studies have demonstrated that the majority of RNA polymerase II (pol II) transcriptional activity takes place outside of protein coding genes, producing numerous new species of ncRNAs from compartments of the genome previously thought to be transcriptionally silent, including intergenic regions and even more strikingly,

RNA Exosome, edited by Torben Heick Jensen.

heterochromatin. These newly discovered RNAs are kept at low levels in the cell and have evaded detection until the development of modern, highly sensitive microarray and sequencing technologies. This raises the important question of how ncRNAs are regulated. The major cellular RNA degrading/processing machinery is the evolutionarily conserved exosome complex, an assembly of 10 protein subunits. The exosomes' 3′ → 5′ exonucleolytic and endonucleolytic activities play extremely versatile functions in the cell, such as regulating the stability of a variety of cellular RNAs, participating in the destruction of viral RNAs,[4-8] 3′-processing of stable structured RNAs and providing quality control for RNA processing, folding and assembly into ribonuclear protein complexes (RNP).[9] Recent data have also implicated the exosome complex to be involved in various aspects of ncRNA metabolism, as well as contributing to gene silencing.[10-12] The analysis of yeast, fly, plant and human transcriptomes revealed dramatic changes of gene expression profiles in exosome mutants.[2,3,5,10,11,13-16] In fission yeast, mutations of exosome subunits lead to the loss of transcriptional silencing from centromeric, telomeric and mating type loci and caused changes in the chromatin modification pattern suggesting that the exosome might be involved in the assembly of silent chromatin.[12,17-20] These data imply that posttranscriptional RNA degradation is an important step in regulating the eukaryotic transcriptome. Here, we focus on recent studies that have unveiled the critical role of the exosome complex in silencing gene expression in different organisms. We compare various aspects of the exosome-mediated silencing pathway to the well studied RNA interference (RNAi) pathway involved in gene silencing and heterochromatin formation and discuss how these pathways cooperate to achieve efficient silencing of gene expression.

HETEROCHROMATIN

In eukaryotes genomic DNA is assembled with histone proteins into nucleosomal structures, forming chromatin. Chromatin remodeling factors along with DNA and histone modifying enzymes are invloved in transcriptional regulation, as well as other DNA transactions such as DNA replication and DNA repair. These activities can alter chromatin structure and therefore make it more accessible for the transcription machinery. Historically, heterochromatin and euchromatin were distinguished as two different genomic compartments, based on the degree of chromatin compaction.[21,22] In contrast to transcriptionally active euchromatin, heterochromatin was therefore believed to be inaccessible for transcription. Heterochromatin is an epigenetically heritable and conserved state of eukaryotic chromosomes and has important functions in protecting genomes from transposable elements, maintaining genome integrity and regulating gene expression. It is typically formed around repetitive DNA sequences, such as those originating from transposable elements, or genes that are found in multiple copies in the genome. Thus, heterochromatin is associated with telomeric, centromeric and ribosomal DNA (rDNA) arrays, together with silenced mating type loci in fungi. Furthermore, the process of gene dosage compensation in mammals involves heterochromatic inactivation of the entire X chromosome in somatic cells. A typical feature of heterochromatin shared by all eukaryotes is hypoacetylation of histones, which is achieved through the enzymatic activity of a conserved NAD^+-dependent histone deacetylase (HDAC) known as Sir2 (*S*ilent *I*nformation *R*egulator) in the budding yeast *Saccharomyces cerevisiae*. In fission yeast and metazoans, this process requires additional HDAC activities. Hypoacetylated histones form a more compact chromatin structure with DNA that was believed to block the recruitment of pol II to DNA (Fig. 1). Indeed, for a long

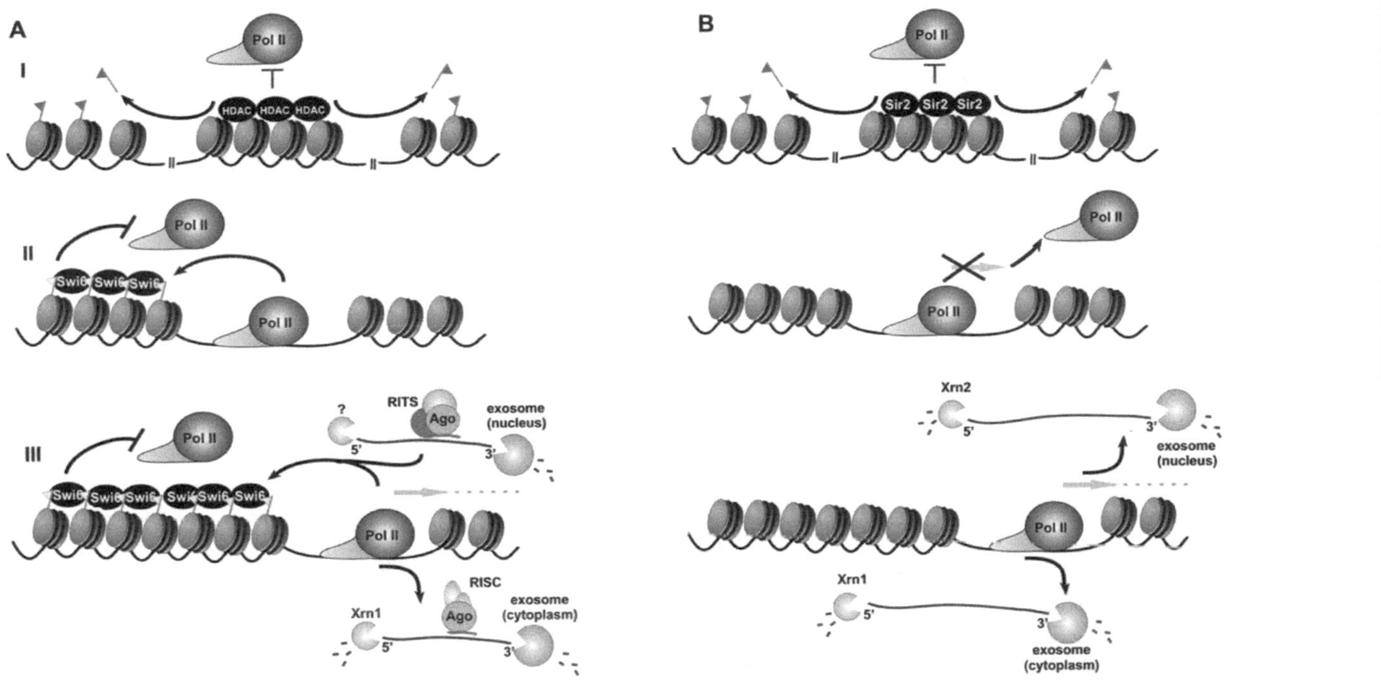

Figure 1. Model for multilevel repression of heterochromatic transcription in RNAi (represented by *S. pombe*) (A) and nonRNAi (represented by *S. cerevisiae*) (B) species. I) Chromatin modifications. The formation of heterochromatin is stimulated by the removal of acetyl marks (red flags) from histones H3 and H4, resulting in the establishment of a more compact structure nonconducive for access by RNA polymerase II (pol II). In *S. cerevisiae*, Sir2 is the major histone deacetylase, in contrast to *S. pombe* and higher eukaryotes that have additional HDACs contributing to histone deacetylation within heterochromatin. II) Regulation of silencing at the transcriptional level. Pol II complexes initiated within heterochromatin can be paused or terminated prematurely. In *S. pombe*, pol II transcription is necessary for the establishment of heterochromatin. siRNA-mediated transcriptional gene silencing (TGS) facilitates methylation of histone tails (yellow flags). Binding of the HP1 homologue Swi6 to methylated histones results in spreading of heterochromatin. III) Degradation of heterochromatic transcripts further contributes to repression. Heterochromatic RNAs can be degraded from their 5′ ends by the exonucleolytic activities of Xrn1 and Xrn2, or from their 3′ ends by the exosome. A color version of this image is available at www.landesbioscience.com/curie.

time the Sir2 dependent mechanism was thought to be sufficient for complete silencing of pol II transcription within the heterochromatin of budding yeast. However, recent studies challenged this notion by demonstrating that although Sir2 reduces the levels of pol II recruitment, heterochromatin remains accessible for pol II, suggesting that silencing is also regulated downstream of pre-initiation complex (PIC) formation.[10,23-25] Indeed, in yeast and flies gene silencing has been shown to occur after PIC formation.[24,26,27] Furthermore, promoter-proximal pausing of pol II seems to be a general feature of eukaryotic promoters, which perhaps reflects an important regulatory stage during pol II transcription.[28,29] The presence of pol II within heterochromatin whilst no detectable RNA is produced, suggested that other mechanisms could contribute to heterochromatic silencing: First, early pol II transcription inhibition/termination and second, RNA degradation. Depletion or mutations of the Nrd1/Nab3/Sen1 complex or exosome subunits (Rrp6 and Rrp4) recently revealed the presence of heterochromatic RNA species.[10] Disruption of the Nrd1-exosome pathway also leads to de-repression of reporter genes integrated into the rDNA array or telomeric regions.[10] Nrd1 associates with transcription complexes proximal to the promoter region by interacting with the phosphorylated form of early elongating pol II.[30,31] This allows Nrd1 to control transcription early in elongation by promoting pol II termination.[32] The Nrd1 complex interacts with the exosome and promotes RNA degradation or processing by facilitating exosome recruitment to the 3′-ends of RNA substrates.[33] Similar effects were also seen when Trf4, a poly(A)polymerase component of the TRAMP complex and a cofactor of the nuclear exosome,[34,35] was deleted.[11] Therefore, heterochromatic silencing in budding yeast is regulated at multiple levels including Sir2-mediated chromatin repression, transcriptional repression and degradation of heterochromatic RNAs.

The proposed multilevel repression system in *S. cerevisiae* has interesting parallels to heterochromatic repression mediated by the RNAi machinery, the mechanism by which this is achieved will be discussed later (Fig. 1). Recent data suggests that in the fission yeast *Schizosaccharomyces pombe*, both heterochromatin formation and the subsequent degradation of nascent heterochromatic transcripts is carried out by the RNAi and exosome machineries.[12,17-20,36,37] Both of these pathways appear to require the spliceosome for the formation of pericentric heterochromatin.[17,38] The RNAi pathway stimulates heterochromatin formation by promoting the establishment of tri-methylation on lysine 9 of histone H3 (H3K9me3), a hallmark of heterochromatin in fission yeast and higher eukaryotes through recruitment of the histone methyltransferase (HMT) Clr4.[39] The H3K9me3 mark provides a high affinity-binding site for the recruitment of Swi6 in *S. pombe* or heterochromatin protein 1 (HP1) in metazoans and allows for the spreading of heterochromatin. Additionally, heterochromatic domains in plants and mammals are known to be methylated through the action of DNA methyltransferases, further contributing to the repression of transcription.

GENE SILENCING PATHWAYS: RNAi AND THE EXOSOME

RNAi is one of the best characterized mechanisms of gene silencing in eukaryotes and is highly conserved. It is essential for protecting cells from invasion by foreign DNA and RNA (transposons, pseudogenes, retroelements and viruses), as well as for the establishment and maintenance of heterochromatin.[40] The RNAi pathway utilizes small ncRNAs, which interact with target RNAs by base pairing and thereby achieve silencing by promoting the degradation of these complementary RNAs and/or translational inhibition (post-transcriptional gene silencing (PTGS)).[41] Moreover, a similar ncRNA-based mechanism mediates transcriptional repression by recruiting chromatin and/or, in some organisms, DNA modifying activities to heterochromatic regions (transcriptional gene silencing (TGS)). Several classes of small ncRNAs have been described in yeast, plants, flies and mammals. Amongst these are small interfering RNAs (siRNAs), microRNAs (miRNAs) and PIWI interacting RNAs (piRNAs). The RNAi pathway is triggered by siRNAs that are generated by the endonuclease Dicer from long double-stranded RNA (dsRNA) precursors which arise as a result of RNA-dependent RNA polymerase (RdRP) activity, transcription of repetitive inverted transposable elements integrated in the genome, or from the dsRNAs of viruses in many organisms that lack RdRP (such as budding yeast, flies and mammals) (Fig. 2). In light of the recent findings that abundant antisense RNAs are produced from both eu- and heterochromatic domains in eukaryotes,[2,3,14,42-45] it is tempting to speculate that antisense RNAs might function by forming duplex RNA thus activating the RNAi pathway. Double stranded, ~22 nucleotide long, siRNAs produced by Dicer cleavage are loaded onto Argonaute containing complexes, the RNA-induced transcriptional silencing complex (RITS) that regulates heterochromatin in the nucleus during TGS, or the RNA-induced silencing complex (RISC) that mediates PTGS in the cytoplasm, both of which direct the cleavage of target RNA.

Although RNAi is an essential pathway in many organisms, it has been lost in some unicellular eukaryotes with small genomes, such as *S. cerevisiae*, *Trypanosoma cruzi*, *Leishmania major*, *Cyanidioschyzon merolae* and *Paramecium falciparum*.[40] However, functional RNAi has been retained in other species closely related to *S. cerevisiae*, such as *Saccharomyces castellii*, *Kluyveromyces polysporus* and *Candida albicans* and can be reconstructed in *S. cerevisiae* if the major RNAi components from *S. casteli*, Dicer (Dcr1) and Argonaute (Ago1), are re-introduced.[46] Analysis of the *S. cerevisiae* transcriptome has revealed that expression of *DCR1* leads to an accumulation of retrotransposon Ty1 siRNA and reduction in the level of corresponding Ty Gag protein and mRNA. These data suggest that a major function of the RNAi pathway is to mediate the silencing of transposable elements; conversely, the lack of RNAi in *S. cerevisiae* suggests that this species relies on alternative pathway(s) to suppress expression of retrotransposons. Importantly, the alternative pathway(s) present in *S. cerevisiae* must be very effective not only because it allowed the loss of RNAi but also because it keeps Ty1 copy number very low (3.1% of the genome).[47] In *S. cerevisiae* Ty silencing is mediated *in trans* by an unstable antisense Ty1 RNA. Over-expression of this antisense RNA from a plasmid, or stabilization of the antisense RNA upon deletion of cellular 5′ → 3′ (*XRN1*) or 3′ → 5′ (*RRP6* and *TRF4*) exonucleolytic activities results in decreased levels of Ty1 RNA. The antisense RNA from the Ty1 element has also been proposed to be part of a posttranslational mechanism that limits retro-transposition by reducing the level of proteins required for replication and integration.[48]

The function of the RNAi pathway in silencing transposable elements is conserved amongst species. The presence of an efficient silencing system may be even more important for multicellular organisms, considering that 42% of the human genome consists of retrotransposons.[49] In *Drosophila* and other multicellular eukaryotes, a piRNA-based mechanism defends the germline against transposons. This surveillance system plays an important role in development, since if transposition occurs within the genome of germline cells, transposons will be passed to the next generation. Interestingly, plants also make use of the exosome complex that operates as a part of the siRNA pathway[50] to silence transposable elements.[2] Tiling microarray experiments in *Arabidopsis thaliana* revealed

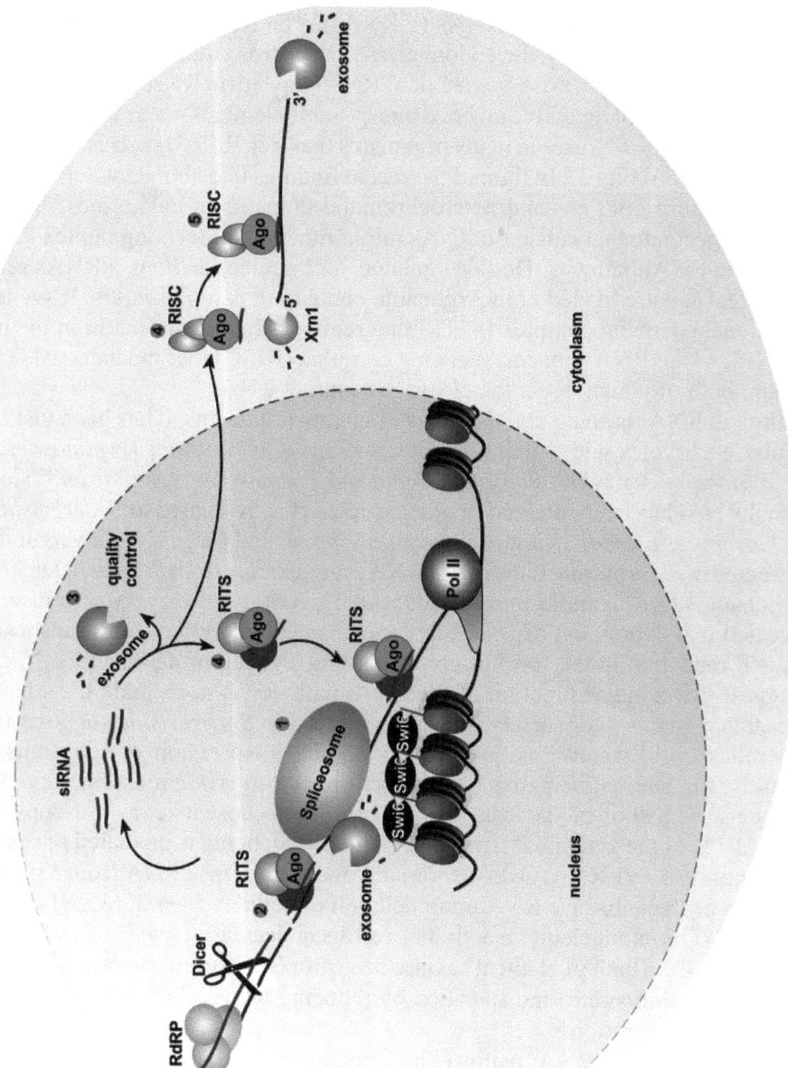

Figure 2. Please see legend on following page.

Figure 2, viewed on previous page. Model illustrating roles of the exosome complex in the RNAi pathway based on the data compiled from various organisms. Generic names of proteins and complexes are used. 1) During the process of cotranscriptional silencing, the RITS complex is recruited to nascent heterochromatic ncRNAs and contributes to the deposition of methyl marks (yellow flags) on histone tails, which facilitates the formation of heterochromatin. In cooperation with the spliceosome, the exosome also facilitates the establishment of H3K9 methylation. HP1 proteins such as Swi6 specifically associate with methylated histones and facilitate the spread of heterochromatin. 2) RNA—dependent RNA polymerase (RdRP) activity synthesizes complementary RNAs resulting in the production of dsRNA, which is cleaved into siRNAs by Dicer. 3) The quality of siRNAs is monitored by the exosome complex and aberrant RNAs are selectively degraded. 4) siRNAs are loaded onto the Argonaute (Ago) containing RITS complex in the nucleus, or RISC complex in the cytoplasm, which can recognize and stimulate the degradation of specific complementary RNAs. 5) In the cytoplasm, the exosome and Xrn1 also mediate the degradation of RISC targeted RNAs by their respective 3′ to 5′ and 5′ to 3′ exonuclease activities. A color version of this image is available at www.landesbioscience.com/curie.

an accumulation of complementary pairs of sense and antisense transcripts, precursors for the ~24 nt endogenous heterochromatic siRNAs, in RRP4- and RRP41-depleted cells. These siRNAs guide repressive modifications (methylation) of DNA and histones at transposon containing loci.

While the function of RNAi as an important cellular defense mechanism is well established[51] and efficiently used in therapeutic applications to target viral RNAs,[52] diverse roles of the exosome machinery during the cellular response to viral invasion have only recently begun to emerge. The nuclear exosome complex down-regulates transcripts encoded by the human endogenous retrovirus (HERV), expressed from within the human genome.[4,5] Furthermore, the exosome also degrades viral mRNA upon infection by RNA viruses such as Moloney murine leukemia virus (MLV) and Sindbis virus (SIN). This exosome-mediated viral silencing is regulated by the Zinc-finger antiviral protein (ZAP), a host antiviral factor that specifically binds to its target mRNAs and recruits the exosome for degradation.[6] Cells therefore utilize several redundant mechanisms (RNAi and the exosome) to efficiently silence transposons and viruses and prevent their proliferation in the cell. However, in some instances an invading virus has been shown to hijack the exosome machinery as part of the disease course. For instance, the virus-encoded SOX (*S*hut*O*ff and e*X*onuclease) protein is proposed to regulate exosome function following infection by Karposi's sarcoma-associated herpesvirus (KSHV)[7] and the exosome core protein Rrp46, which mediates RNA processing and degradation, is involved in the initiation of programmed cell death caused by a virus in barley.[8]

In addition to its function in silencing transposons, RNAi promotes the establishment and spreading of heterochromatin.[53] Similar to the regulation of protein coding genes, heterochromatic silencing is achieved as a result of transcriptional and posttranscriptional regulation (Fig. 1). The epigenetic regulation of heterochromatin silencing by RNAi was initially reported in fission yeast, plants and flies and more recently has been suggested to occur in mammals.[54,55] During TGS, the RNAi containing RITS complex targets nascent heterochromatic transcripts and mediates their degradation (Fig. 2).[36] As previously mentioned, this process leads to changes in the methylation status of histone tails that subsequently result in transcriptional repression. Strikingly, the spliceosome was recently suggested to play a role in the generation of siRNAs and in the degradation of heterochromatic RNA precursors during TGS.[38] This was proposed to facilitate the amplification of siRNAs to high levels and thus promote effective heterochromatic silencing. Unexpectedly, the RNA duplexes that induce transcriptional silencing can also induce transcriptional activation in human cell lines.[56] The precise mechanism by which the same pool of small RNAs can induce both silencing and activation of transcription is unknown.

The exosome complex also contributes to heterochromatic silencing independently of RNAi.[2,17] One mechanism by which the exosome regulates heterochromatic silencing is by directly degrading RNAs produced from these regions. In *S. cerevisiae*, the exosome degrades heterochromatic transcripts produced by pol II from rDNA and telomeric loci[10,11] and released as a result of transcription termination by the Nrd1 complex. The exosome cofactor Trf4 is also required for the degradation of these cryptic RNAs. Similarly, Cid14, the *S. pombe* homologue of *S. cerevisiae* Trf4, promotes exosome-mediated degradation of heterochromatic transcripts.[12] Furthermore, the exosome also cooperates with components of the spliceosome to degrade heterochromatic RNAs.[17] Dis3, a subunit of the exosome that mediates its endo- and exonucleolytic activities, was shown to colocalize with splicing factor Spf30 and facilitate its recruitment to pericentric heterochromatin. Mutations in Spf30 and Rrp6 lead to reduced H3K9me3 at the centromeres and Dis3 deficiency results in reduced recruitment of Swi6,[19] suggesting that both splicing and exosome complexes may contribute to heterochromatin formation.

Finally, RNAs targeted by the RNAi pathway can be cleaved by RISC and degraded in the cytoplasm, in a process that again converges the RNAi pathway with exosome activity (Fig. 2). In *Drosophila*, 5′ mRNA fragments generated by RISC cleavage are rapidly degraded from their 3′ ends by the exosome, whereas the 3′ fragments are degraded from their 5′ ends by XRN1.[57] Similarly, in the algae *Chlamydomonas reinhardtii*, the exosome is involved in degrading RNA fragments cleaved by the RISC complex;[58] RNAs generated by RISC are polyadenylated by MUT68, which is thought to promote exosome mediated degradation. In conclusion, the exosome plays versatile functions in regulating gene silencing both independently of RNAi and as a part of the RNAi system.

MULTILEVEL REGULATION OF GENE EXPRESSION

As illustrated above a three level model can be used to explain the silencing of heterochromatic transcription in eukaryotes, in which silencing is established at chromatin, transcriptional and posttranscriptional levels (Fig. 1). Although all three levels of repression could function completely independently, the cross-talk between transcriptional and posttranscriptional processes appears to be important to achieve efficient silencing. Indeed, on-going pol II transcription is required for efficient RITS mediated heterochromatic silencing[36] and the exosome complex is physically and functionally linked to transcription.[17,33,59] The RNAi pathway is proposed to operate in a reverse feedback loop, whereby it regulates chromatin by regulating the processing of heterochromatic precursor RNAs (Fig. 2).

The multilevel repression of heterochromatin is not limited to RNAi containing species, *S. cerevisiae* also uses multilevel mechanisms to repress pol II transcription within heterochromatin. Complete repression is achieved by the combined actions of Sir2, the Nrd1 transcription terminator complex and the exosome. The Nrd1 complex mediates termination of transcripts such as short (about ≤500 nt) mRNAs, snoRNAs and a recently described new class of ncRNAs-Cryptic Unstable Transcripts (CUTs). CUTs are widespread throughout both heterochromatic and euchromatic domains of the eukaryotic genome and are usually degraded by the exosome.[3,5,13,60-63] However, recent studies have shown that transcription of noncoding regions can have important regulatory functions.

They demonstrate that the act of noncoding transcription per se can affect interactions between nearby DNA and histones or other DNA binding proteins and can thus regulate the expression of neighboring genes. Indeed, the failure to properly terminate CUT transcription leads to dramatic changes in chromatin structure within the rDNA region, most notably an increase in histone acetylation, suggesting that the role of pol II nc transcription within the rDNA array could be to regulate the transcription of ribosomal genes.[10] This seems to be an important general mechanism for regulating gene expression, since a number of genes have been shown to be regulated according to this scenario. For example, transcription of ncRNA is required for chromatin remodeling and activation of the *fbp1*$^+$ locus.[64] This was convincingly demonstrated by inserting a transcription terminator upstream of the *fbp1*$^+$ gene that prevented formation of an open chromatin structure and lead to transcriptional repression of the gene. Similarly, antisense transcription at the *PHO5* promoter was shown to activate transcription of the *PHO5* gene.[65] Conversely, nc transcription can negatively affect the transcription of neighboring genes as was reported for the *SER3* and *FLO11* genes.[66,67] In *Arabidopsis*, transcription of the intergenic region by RNA polymerase IVb/Pol V, required for siRNA-mediated gene silencing of transposons, results in transcriptional silencing of overlapping and adjacent genes.[68]

Interestingly, genetic evidence further supports the notion of a functional connection between RNA turn-over and chromatin structure.[69,70] The exosome cofactors Nrd1 and Trf4 show strong negative genetic interactions with the components of the Rpd3L HDAC complex (Pho23, Sap30, Sds3, Rxt2 and Sin3)[69,70] and with Set1 methyltransferase, which introduces H3K4me marks, suggesting a functional connection between these genes (Terzi and Buratowski, personal communication). Additionally, strains combining the deletion of *RRP6* and *TRF4* genes with deletions of Set1 complex components (*SDC1* and *SWD1*) also exhibit a synthetic sick phenotype.[71] Furthermore, in budding yeast Set1 HMT activity is known to contribute to heterochromatic silencing, in addition to Sir2.[72] Similarity in the distribution pattern of H3K4me3 and Nrd1 localization at the 5′ end of genes implies that these two proteins may cooperate in regulating early events during pol II transcription. Indeed, further analyses revealed that Set1 and the Rpd3L complex acting downstream of Set1, facilitate the recruitment of Nrd1 and thereby contribute to Nrd1 mediated gene silencing within heterochromatin. These data may help to explain earlier studies that implicated a role for Set1 in transcriptional repression within heterochromatin,[72-74] which were somewhat contradictory to the well established role that Set1 plays in activating euchromatic transcription[75,76] Interestingly, in humans H3K4me helps to recruit components of the spliceosome[77] and in *S. pombe*, the spliceosome was shown to be involved in the establishment and maintenance of heterochromatic chromatin.[17,38] Together with the evolutionary conservation of H3K4me these observations suggest an intriguing possibility that cross-talk between chromatin and RNA processing pathways is a common theme used by eukaryotes to control gene silencing.

Although RNAi has been shown to regulate gene silencing in fission yeast, RNAi-independent pathways are also used. As in *S. cerevisiae*, these pathways also rely on the exosome.[12,17-20,37,78] Consistent with its role in the destruction of spurious ncRNAs, combining the deletion of *rrp6* with mutations in *CLR6* (HDAC) in *S. pombe* leads to a synergistic increase in antisense transcription.[20] Additionally, HIRA (replication independent histone chaperone complex) components, involved in transcriptional silencing at heterochromatic loci[79] and suppression of antisense transcription, show synthetic lethality when combined with Δ*rrp6*.[8] Furthermore, the conserved variant histone H2A.Z (*PTH1* gene) that regulates heterochromatic silencing at centromeres and spurious intergenic transcription in fission yeast[37,80] is loaded onto the 5′ ends of genes by the Swr1 complex, in cooperation with either RNAi or exosome components, to suppress antisense transcription.[37] In *S. cerevisiae*, H2A.Z is also implicated in preventing the spread of silent heterochromatin.[81] Similar to its homologues from *S. pombe* and other eukaryotes,[82] *S. cerevisiae* H2A.Z is found at the 5′ end of genes,[83] indicating that its' suppression of antisense transcription may be an evolutionarily conserved mechanism. Furthermore, double mutants lacking Htz1 (the gene encoding for H2A.Z in *S. cerevisiae*) and the exosome cofactors Trf4[84,85] or Pab1[71,86] show a synthetic growth defect. A similar growth defect was observed for double mutants lacking Nrd1, Trf4 and the exosome subunit Rrp6 together with components of the SWR1 complex (Swr1, Swc3, Swc5 and Swc6), involved in deposition of H2A.Z.[69,70] Together, these data strongly support the proposed cross-talk between RNA turn-over and chromatin.

Finally, recent studies have established ncRNAs themselves as regulators of chromatin. The function of the exosome in regulating the expression of regulatory ncRNAs thus introduces another level to the regulation of gene silencing which is discussed below.

REGULATING THE REGULATORS: HOW THE STABILITY OF ncRNAs DETERMINES THEIR FUNCTION IN HETEROCHROMATIN FORMATION

ncRNAs show no obvious sequence or structural conservation and vary largely between different organisms. However, a common theme emerging from the model systems analyzed to date is their role in regulating gene expression at the level of chromatin. Classical examples include *Xist*, *Tsix* and *roX* RNAs involved in dosage compensation, and *Air* and *H19* RNAs involved in genomic imprinting. Over recent years, the repertoire of ncRNAs has expanded, as efforts to look into the transcriptional output of eukaryotic genomes have increased. As previously mentioned, most of these newly found ncRNAs are highly unstable, posing an interesting question as to whether the RNA produced is functional, or the result of transcriptional noise.[87] Accumulating evidence implicates many ncRNAs in regulating gene silencing. Interestingly, their function in the cell can be dynamically controlled by the exosome complex, which maintains appropriate levels of ncRNAs in the cell, thus implicating the exosome as a regulator of gene silencing.

Recently, mammalian and yeast telomeres were found to be actively transcribed into telomeric repeat-containing RNA (TERRA) molecules.[88-91] TERRA associates with telomeric regions of metaphase chromosomes in cis to provoke telomere shortening and genomic instability. This appears to be achieved by two mechanisms. First, it was demonstrated biochemically that TERRA inhibits telomerase, a reverse transcriptase that can add telomeric repeats onto chromosome ends.[92] Other studies later demonstrated that TERRA is involved in the establishment of H3K9me3, an epigenetic modification typical of highly compacted heterochromatin and a common feature of telomeric repeats in all organisms.[93,94] H3K9me3 plays an important role in telomere-length homeostasis,[90] indeed, mouse loss of function studies have revealed that impairment of the telomeric heterochromatin structure alters telomere length.[92] Importantly, studies using immortalized mouse cell lines showed that the average telomere length correlates with TERRA levels, implying that maintaining an appropriate amount of TERRA in the cell is important for its function.[92] Interestingly, earlier studies from the Gottschling laboratory established an initial connection between cellular levels of *TLC1* RNA (the RNA component of the telomerase complex in *S. cerevisiae*) and telomeric silencing.[95,96] These studies demonstrated that plasmid driven over-expression of *TLC1* RNA results in a dramatic loss of telomeric silencing. With such key roles in regulating heterochromatic silencing and chromosome stability, it is critical that the levels of TERRA and telomerase RNAs are appropriately controlled. Interestingly, the effectors of nonsense-mediated messenger RNA decay, SMG proteins (Suppressors with Morphogenetic defects in Genitalia) were shown to modulate the actions of TERRA and telomerase RNA molecules during telomere length regulation by affecting their levels and cellular localization. SMG1 regulates TERRA function by promoting its displacement from telomeric chromatin,[88] whilst other SMG proteins UPF1 (SMG2) and EST1A (SMG6), associate with telomeric chromatin and physically interact with the RNA component of telomerase, contributing to its activity in replicating telomere ends.[97-99] Furthermore, in *S. cerevisiae*, the nuclear $5' \rightarrow 3'$ exonuclease Rat1 degrades TERRA RNA and thus promotes telomere elongation.[89] Telomeric RNA levels are also regulated by the cytoplasmic $5' \rightarrow 3'$ exonuclease Xrn1 and the cofactor of the nuclear exosome complex Trf4.[100] Moreover, exosome mutants display an accumulation of *TLC1* RNA,[101] (Coy and Vasiljeva, unpublished) and its polyadenylated precursor,[102] suggesting that the exosome may be involved in regulating *TLC1* RNA levels and processing. In support of this view, mutations of the exosome subunits Rrp6 and Rrp47 lead to a shortening of telomere ends.[101,103] Together, these studies represent an example of a molecular link between RNA regulation and the maintenance of telomere length homeostasis. Furthermore, a global role for the exosome complex in regulating the stability of antisense RNAs is also becoming apparent. Recent studies in a variety of organisms have demonstrated that the vast number of antisense transcripts are stabilized upon inactivation of the exosome.[2,3,5,13-16] This phenomenon is exemplified by the antisense Ty1 RNA[100,104] and antisense *PHO84* RNA, both of which direct the silencing of transcription from the opposite strand in *S. cerevisiae*.[105,106] In each case, stabilization of antisense RNAs is a key event in the silencing pathway. The stabilization of Ty1 antisense RNA observed in strains harboring a deletion of the cytoplasmic exonuclease Xrn1, caused *trans*-silencing of Ty1 expression. Although primarily regulated by Xrn1, the exosome too participates in controlling the abundance of Ty1 antisense RNA, since deletion of Rrp6 or Trf4 also lead to its accumulation. The stability of antisense *PHO84* RNA is dynamically controlled by the nuclear exosome complex during the process of chronological aging. The antisense RNA accumulates in aging cells, correlating with a decrease in *PHO84* mRNA. According to the model proposed by the authors, antisense *PHO84* RNA directs recruitment of the Hda1 histone deacetylase complex, promoting H3K18 deacetylation and subsequent silencing of sense *PHO84* transcription *in cis* and *in trans*. Given that antisense transcription is widespread in eukaryotic genomes it is tempting to speculate that this could be a general mechanism contributing to gene regulation.

Unlike most of the newly found ncRNAs, *Xist* and its antisense *Tsix* are relatively stable RNAs (in undifferentiated embryonic stem cells *Xist* half-life ($t_{1/2}$) is 3-6hrs, *Tsix* $t_{1/2}$ is <1hr).[107] *Xist* is known to silence transcription from the inactive X chromosome in mammals while *Tsix* is believed to function as a negative regulator of *Xist*.[108] *Xist* promotes heterochromatization *in cis* by coating the X chromosome, a process that coincides with an accumulation of transcriptionally repressive H3K27me3 and ubiquitination of lysine 119 on histone H2A (uH2A) marks, mediated by polycomb group proteins.[109,110] Strikingly, down-regulation of *Exosc10*, the human homologue of Rrp6 and *Rent1*, the homologue of Upf1, leads to subsequent down-regulation of *Xist* RNA and inhibition of the onset of X-inactivation.[111] It remains to be seen whether this effect is direct; the decrease in *Xist* RNA levels, as opposed to the accumulation of RNA normally associated with exosome mutants implies that the exosome is not directly involved in this process. Alternatively, the exosome may be involved in the regulation of *Tsix* RNA, whose expression inversely-correlates with the expression of *Xist*.[112] Recently an intriguing link between X chromosome inactivation and the RNAi pathway was also found. The complementary *Xist* and *Tsix* RNAs were shown to form duplexes that are processed into siRNA in a Dicer-dependent manner. These Dicer generated siRNAs are thought to be important for the accumulation of H3K27me3 and for X chromosome silencing.[113] However, results of Nesterova et al[114] argued against a direct role for RNAi in X chromosome inactivation by showing that upon depletion of DNA histone methyltrasferases in Dicer-deficient male ES cell lines, despite an increase in Xist expression, X chromosome silencing still occurs. Interpretation of the results is complicated by the death of Dicer mutants shortly after implantation, leaving a very short window in which to assay for chromosome inactivation, but clearly additional studies are needed to determine what role if any RNAi plays in the process of X chromosome inactivation.

In contrast to mammals, in *Drosophila* transcription from the single X chromosome in males is up-regulated to compensate for two transcriptionally active X chromosomes in females.[115] The MSL complex, an assembly of several proteins and two ncRNAs, *roX1* and *roX2* (RNA on the X), encoded by the male specific lethal genes is involved in up-regulating the expression of genes on the male X chromosome. Currently, the mechanism by which the MSL RNA-protein complex acts to regulate dosage compensation in flies is not understood. Interestingly, the MSL complex copurifies with components of the exosome.[116] It has been speculated that recruitment of the nuclear exosome complex to transcription sites regulated by the MSL complex is necessary to prevent excessive production of mRNAs and to destroy cryptic transcripts that may result from open chromatin.

The levels and quality of siRNAs and miRNAs can also be regulated in the cell. In *S. pombe* the conserved exonuclease Eri1 specifically degrades siRNA duplexes. Upon deletion of Eri1 there is an increase in the proportion of siRNAs associated with the RITS complex, corresponding to an increase in heterochromatic silencing that is accompanied by increased levels of H3K9me3.[117] Along with the majority of other cellular RNAs, the exosome also monitors quality of siRNAs and dysfunctional RNA molecules are destroyed.[58,118] Similar to other exosome substrates, siRNAs that are selected for exosome mediated degradation are marked by either poly(A) or poly(U) tails, which appears to be a conserved mechanism in RNAi containing organisms as well in *S. cerevisiae*. In the algae *Chlamidomonas*, the terminal transferase MUT68 directs untemplated uridylation of siRNAs and thus stimulates their degradation by the exosome.[118] It was speculated that the exosome monitors the quality of these small RNAs and destroys dysfunctional molecules (Fig. 2). The same enzyme can also catalyze addition of oligo(A) tails to the 3′ end of RISC generated RNAs that are degraded by the exosome.[58] Furthermore, in *Caenorhabditis elegans* a new class of small RNAs (22G-RNAs) associating with two distinct Argonaute proteins, (WAGOs) and CSR-1, has recently been implicated in the silencing of transposons and repetitive elements, as well as chromosome segregation. These small RNAs are also oligouridylated and in the absence of CDE-1 poly(U) polymerase they accumulate to inappropriate levels. These data suggest that 22G-RNAs may be targeted by the exosome complex for degradation. Elevated 22G-RNA levels are associated with erroneous gene silencing, most likely through the inappropriate loading of siRNAs onto other Argonaute proteins, resulting in defects in both meiotic and mitotic chromosome segregation.[119]

CONCLUSION

It is becoming increasingly clear that the regulation of gene silencing is a complex process composed of multiple interconnected steps. Traditionally, gene regulation has been thought to occur at the chromatin level through regulating the access of polymerases. It now transpires that gene silencing is controlled at multiple levels, also encompassing the regulation of transcription and RNA stability. The cross-talk between different levels of regulation is a well-conserved phenomenon of eukaryotes and may have been crucial in allowing the evolution of complex organisms.

The exosome, with its critical roles in regulating the abundance of RNA transcripts, has been unveiled as a key component in the story of gene regulation. In addition to protecting the genome against viral invasion, the exosome contributes to heterochromatic silencing by degrading RNA products from heterochromatic loci. In this role the exosome acts independently and is partially redundant with the RNAi machinery also capable of silencing heterochromatic RNA transcripts via the RITS complex. Interestingly, whilst some unicellular organisms have lost components of the RNAi pathway, an exosome complex is conserved in all phyla. It is therefore likely that the exosome existed to regulate gene expression before the evolution of the RNAi system. Indeed, as the complexity of eukaryotic genomes increased, so too did the proportion of noncoding RNAs expressed, correlating with increased demand for RNA degradation machineries. The redundancy between different gene silencing mechanisms could be important in the constant battle faced by the cell in policing the transcriptome, since the additional monitoring ensures that any errors are rapidly recognized and effectively suppressed, thus increasing the fidelity of transcription.

The emergence of regulatory ncRNAs as functional entities in their own right, together with the discovery that RNA synthesis is far more abundant than originally thought, have challenged our understanding of genome regulation. Non-coding RNAs have emerged as important regulators of genome silencing both at the transcriptional and posttranscriptional level. Consequently, the exosome and other RNA degrading enzymes have been shown to play critical roles in regulating ncRNA metabolism. Thus, the exosome is transpiring to be a versatile machine capable of recognizing a range of substrates. It seems that the flexibility of the exosome has been harnessed by the cell to allow the high levels of transcription recently exposed in eukaryotic genomes. Future challenges will focus on establishing the specific roles of these RNAs, how their expression is regulated and exactly how their interaction at the chromatin level is mediated. One thing is clear: the complexity of gene expression will continue to fascinate us for quite sometime to come.

ACKNOWLEDGEMENTS

We apologize to those colleagues whose important contribution to the field was not cited due to space limitations. We thank N. Terzi and S. Buratowski for sharing data prior to publication. We also thank M. Bühler, M. Cohn, T. Nesterova and N. Terzi for critical reading of manuscript. We are very grateful to our colleagues at the Department of Biochemistry, University of Oxford for helpful discussions, encouragement and support. LV research is supported by the Leukemia and Lymphoma Society transition grant and a Career Development Award from Wellcome Trust.

REFERENCES

1. Birney E, Stamatoyannopoulos JA, Dutta A et al. Identification and analysis of functional elements in 1% of the human genome by the ENCODE pilot project. Nature 2007; 447(7146):799-816.
2. Chekanova JA, Gregory BD, Reverdatto SV et al. Genome-wide high-resolution mapping of exosome substrates reveals hidden features in the Arabidopsis transcriptome. Cell 2007; 131(7):1340-53.
3. Wilhelm BT, Marguerat S, Watt S et al. Dynamic repertoire of a eukaryotic transcriptome surveyed at single-nucleotide resolution. Nature 2008; 453(7199):1239-43.
4. Kammler S, Lykke-Andersen S, Jensen TH. The RNA exosome component hRrp6 is a target for 5-fluorouracil in human cells. Mol Cancer Res 2008; 6(6):990-5.
5. Preker P, Nielsen J, Kammler S et al. RNA exosome depletion reveals transcription upstream of active human promoters. Science 2008; 322(5909):1851-4.
6. Guo X, Ma J, Sun J et al. The zinc-finger antiviral protein recruits the RNA processing exosome to degrade the target mRNA. Proc Natl Acad Sci USA 2007; 104(1):151-6.
7. Glaunsinger BA, Ganem DE. Messenger RNA turnover and its regulation in herpesviral infection. Adv Virus Res 2006; 66:337-94.
8. Anderson HE, Wardle J, Korkut SV et al. The fission yeast HIRA histone chaperone is required for promoter silencing and the suppression of cryptic antisense transcripts. Mol Cell Biol 2009; 29(18):5158-67.
9. Villa T, Rougemaille M, Libri D. Nuclear quality control of RNA polymerase II ribonucleoproteins in yeast: tilting the balance to shape the transcriptome. Biochim Biophys Acta 2008; 1779(9):524-31.
10. Vasiljeva L, Kim M, Terzi N et al. Transcription termination and RNA degradation contribute to silencing of RNA polymerase II transcription within heterochromatin. Mol Cell 2008; 29(3):313-23.
11. Houseley J, Kotovic K, El Hage A et al. Trf4 targets ncRNAs from telomeric and rDNA spacer regions and functions in rDNA copy number control. EMBO 2007; 26(24):4996-5006.
12. Buhler M, Haas W, Gygi SP et al. RNAi-dependent and -independent RNA turnover mechanisms contribute to heterochromatic gene silencing. Cell 2007; 129(4):707-21.
13. Neil H, Malabat C, d'Aubenton-Carafa Y et al. Widespread bidirectional promoters are the major source of cryptic transcripts in yeast. Nature 2009; 457(7232):1038-42.
14. Xu Z, Wei W, Gagneur J et al. Bidirectional promoters generate pervasive transcription in yeast. Nature 2009; 457(7232):1033-7.
15. Davis CA, Ares M Jr. Accumulation of unstable promoter-associated transcripts upon loss of the nuclear exosome subunit Rrp6p in Saccharomyces cerevisiae. Proc Natl Acad Sci USA 2006; 103(9):3262-7.
16. Kiss DL, Andrulis ED. Genome-wide analysis reveals distinct substrate specificities of Rrp6, Dis3 and core exosome subunits. RNA.
17. Bernard P, Drogat J, Dheur S et al. Splicing factor spf30 assists exosome-mediated gene silencing in fission yeast. Mol Cell Biol 30(5):1145-57.
18. Wang SW, Stevenson AL, Kearsey SE et al. Global role for polyadenylation-assisted nuclear RNA degradation in posttranscriptional gene silencing. Mol Cell Biol 2008; 28(2):656-65.
19. Murakami H, Goto DB, Toda T et al. Ribonuclease activity of Dis3 is required for mitotic progression and provides a possible link between heterochromatin and kinetochore function. PLoS ONE 2007; 2:e317.
20. Nicolas E, Yamada T, Cam HP et al. Distinct roles of HDAC complexes in promoter silencing, antisense suppression and DNA damage protection. Nat Struct Mol Biol 2007; 14(5):372-80.
21. Grewal SI, Jia S. Heterochromatin revisited. Nat Rev Genet 2007; 8(1):35-46.
22. Grewal SI, Moazed D. Heterochromatin and epigenetic control of gene expression. Science 2003; 301(5634):798-802.
23. Gao L, Gross DS. Using genomics and proteomics to investigate mechanisms of transcriptional silencing in Saccharomyces cerevisiae. Brief Funct Genomic Proteomic 2006; 5(4):280-8.
24. Sekinger EA, Gross DS. Silenced chromatin is permissive to activator binding and PIC recruitment. Cell 2001; 105(3):403-14.
25. Steinmetz EJ, Warren CL, Kuehner JN et al. Genome-wide distribution of yeast RNA polymerase II and its control by Sen1 helicase. Mol Cell 2006; 24(5):735-46.
26. Dellino GI, Schwartz YB, Farkas G et al. Polycomb silencing blocks transcription initiation. Mol Cell 2004; 13(6):887-93.
27. Schwartz YB, Kahn TG, Dellino GI et al. Polycomb silencing mechanisms in Drosophila. Cold Spring Harb Symp Quant Biol 2004; 69:301-8.
28. Core LJ, Lis JT. Transcription regulation through promoter-proximal pausing of RNA polymerase II. Science 2008; 319(5871):1791-2.
29. Core LJ, Waterfall JJ, Lis JT. Nascent RNA sequencing reveals widespread pausing and divergent initiation at human promoters. Science 2008; 322(5909):1845-8.
30. Gudipati RK, Villa T, Boulay J et al. Phosphorylation of the RNA polymerase II C-terminal domain dictates transcription termination choice. Nat Struct Mol Biol 2008; 15(8):786-94.

31. Vasiljeva L, Kim M, Mutschler H et al. The Nrd1-Nab3-Sen1 termination complex interacts with the Ser5-phosphorylated RNA polymerase II C-terminal domain. Nat Struct Mol Biol 2008; 15(8):795-804.
32. Arigo JT, Carroll KL, Ames JM et al. Regulation of yeast NRD1 expression by premature transcription termination. Mol Cell 2006; 21(5):641-51.
33. Vasiljeva L, Buratowski S. Nrd1 interacts with the nuclear exosome for 3′ processing of RNA polymerase II transcripts. Molecular cell 2006; 21(2):239-48.
34. LaCava J, Houseley J, Saveanu C et al. RNA degradation by the exosome is promoted by a nuclear polyadenylation complex. Cell 2005; 121(5):713-24.
35. Vanacova S, Wolf J, Martin G et al. A new yeast poly(A) polymerase complex involved in RNA quality control. PLoS Biol 2005; 3(6):e189.
36. Buhler M, Verdel A, Moazed D. Tethering RITS to a nascent transcript initiates RNAi- and heterochromatin-dependent gene silencing. Cell 2006; 125(5):873-86.
37. Zofall M, Fischer T, Zhang K et al. Histone H2A.Z cooperates with RNAi and heterochromatin factors to suppress antisense RNAs. Nature 2009; 461(7262):419-22.
38. Bayne EH, Portoso M, Kagansky A et al. Splicing factors facilitate RNAi-directed silencing in fission yeast. Science 2008; 322(5901):602-6.
39. Bayne EH, White SA, Kagansky A et al. Stc1: a critical link between RNAi and chromatin modification required for heterochromatin integrity. Cell 140(5):666-77.
40. Cerutti H, Casas-Mollano JA. On the origin and functions of RNA-mediated silencing: from protists to man. Curr Genet 2006; 50(2):81-99.
41. Malone CD, Hannon GJ. Small RNAs as guardians of the genome. Cell 2009; 136(4):656-68.
42. He Y, Vogelstein B, Velculescu VE, et al. The antisense transcriptomes of human cells. Science 2008; 322(5909):1855-7.
43. Carninci P, Kasukawa T, Katayama S et al. The transcriptional landscape of the mammalian genome. Science 2005; 309(5740):1559-63.
44. Katayama S, Tomaru Y, Kasukawa T et al. Antisense transcription in the mammalian transcriptome. Science 2005; 309(5740):1564-6.
45. David L, Huber W, Granovskaia M et al. A high-resolution map of transcription in the yeast genome. Proc Natl Acad Sci USA 2006; 103(14):5320-5.
46. Drinnenberg IA, Weinberg DE, Xie KT et al. RNAi in budding yeast. Science 2009; 326(5952):544-50.
47. Kim JM, Vanguri S, Boeke JD et al. Transposable elements and genome organization: a comprehensive survey of retrotransposons revealed by the complete Saccharomyces cerevisiae genome sequence. Genome Res 1998; 8(5):464-78.
48. Matsuda E, Garfinkel DJ. Posttranslational interference of Ty1 retrotransposition by antisense RNAs. Proc Natl Acad Sci USA 2009; 106(37):15657-62.
49. Lander ES, Linton LM, Birren B et al. Initial sequencing and analysis of the human genome. Nature 2001; 409(6822):860-921.
50. Alvarado V, Scholthof HB. Plant responses against invasive nucleic acids: RNA silencing and its suppression by plant viral pathogens. Semin Cell Dev Biol 2009; 20(9):1032-40.
51. Obbard DJ, Gordon KH, Buck AH et al. The evolution of RNAi as a defence against viruses and transposable elements. Philos Trans R Soc Lond B Biol Sci 2009; 364(1513):99-115.
52. Lopez-Fraga M, Martinez T, Jimenez A. RNA interference technologies and therapeutics: from basic research to products. BioDrugs 2009; 23(5):305-32.
53. Suzuki K, Kelleher AD. Transcriptional regulation by promoter targeted RNAs. Curr Top Med Chem 2009; 9(12):1079-87.
54. Han J, Kim D, Morris KV. Promoter-associated RNA is required for RNA-directed transcriptional gene silencing in human cells. Proc Natl Acad Sci USA 2007; 104(30):12422-7.
55. Wierzbicki AT, Ream TS, Haag JR et al. RNA polymerase V transcription guides ARGONAUTE4 to chromatin. Nat Genet 2009; 41(5):630-4.
56. Li LC, Okino ST, Zhao H et al. Small dsRNAs induce transcriptional activation in human cells. Proc Natl Acad Sci USA 2006; 103(46):17337-42.
57. Orban TI, Izaurralde E. Decay of mRNAs targeted by RISC requires XRN1, the Ski complex and the exosome. RNA 2005; 11(4):459-69.
58. Ibrahim F, Rohr J, Jeong WJ et al. Untemplated oligoadenylation promotes degradation of RISC-cleaved transcripts. Science 2006; 314(5807):1893.
59. Andrulis ED, Werner J, Nazarian A et al. The RNA processing exosome is linked to elongating RNA polymerase II in Drosophila. Nature 2002; 420(6917):837-41.
60. Arigo JT, Eyler DE, Carroll KL et al. Termination of cryptic unstable transcripts is directed by yeast RNA-binding proteins Nrd1 and Nab3. Mol Cell 2006; 23(6):841-51.

61. Thiebaut M, Kisseleva-Romanova E, Rougemaille M et al. Transcription termination and nuclear degradation of cryptic unstable transcripts: a role for the nrd1-nab3 pathway in genome surveillance. Mol Cell 2006; 23(6):853-64.
62. Kim M, Vasiljeva L, Rando OJ et al. Distinct pathways for snoRNA and mRNA termination. Mol Cell 2006; 24(5):723-34.
63. Steinmetz EJ, Conrad NK, Brow DA et al. RNA-binding protein Nrd1 directs poly(A)-independent 3′-end formation of RNA polymerase II transcripts. Nature 2001; 413(6853):327-31.
64. Hirota K, Miyoshi T, Kugou K et al. Stepwise chromatin remodelling by a cascade of transcription initiation of noncoding RNAs. Nature 2008; 456(7218):130-4.
65. Uhler JP, Hertel C, Svejstrup JQ. A role for noncoding transcription in activation of the yeast PHO5 gene. Proc Natl Acad Sci USA 2007; 104(19):8011-6.
66. Martens JA, Wu PY, Winston F. Regulation of an intergenic transcript controls adjacent gene transcription in Saccharomyces cerevisiae. Genes Dev 2005; 19(22):2695-704.
67. Bumgarner SL, Dowell RD, Grisafi P et al. Toggle involving cis-interfering noncoding RNAs controls variegated gene expression in yeast. Proc Natl Acad Sci USA 2009; 106(43):18321-6.
68. Wierzbicki AT, Haag JR, Pikaard CS. Noncoding transcription by RNA polymerase Pol IVb/Pol V mediates transcriptional silencing of overlapping and adjacent genes. Cell 2008; 135(4):635-48.
69. Collins SR, Miller KM, Maas NL et al. Functional dissection of protein complexes involved in yeast chromosome biology using a genetic interaction map. Nature 2007; 446(7137):806-10.
70. Costanzo M, Baryshnikova A, Bellay J et al. The genetic landscape of a cell. Science 327(5964):425-31.
71. Wilmes GM, Bergkessel M, Bandyopadhyay S et al. A genetic interaction map of RNA-processing factors reveals links between Sem1/Dss1-containing complexes and mRNA export and splicing. Mol Cell 2008; 32(5):735-46.
72. Bryk M, Briggs SD, Strahl BD et al. Evidence that Set1, a factor required for methylation of histone H3, regulates rDNA silencing in S. cerevisiae by a Sir2-independent mechanism. Curr Biol 2002; 12(2):165-70.
73. Briggs SD, Bryk M, Strahl BD et al. Histone H3 lysine 4 methylation is mediated by Set1 and required for cell growth and rDNA silencing in Saccharomyces cerevisiae. Genes Dev 2001; 15(24):3286-95.
74. Pinskaya M, Morillon A. Histone H3 lysine 4 di-methylation: a novel mark for transcriptional fidelity? Epigenetics 2009; 4(5):302-6.
75. Dehe PM, Geli V. The multiple faces of Set1. Biochem Cell Biol 2006; 84(4):536-48.
76. Shilatifard A. Molecular implementation and physiological roles for histone H3 lysine 4 (H3K4) methylation. Curr Opin Cell Biol 2008; 20(3):341-8.
77. Sims RJ 3rd, Millhouse S, Chen CF et al. Recognition of trimethylated histone H3 lysine 4 facilitates the recruitment of transcription postinitiation factors and premRNA splicing. Mol Cell 2007; 28(4):665-76.
78. Yamada T, Fischle W, Sugiyama T et al. The nucleation and maintenance of heterochromatin by a histone deacetylase in fission yeast. Mol Cell 2005; 20(2):173-85.
79. Greenall A, Williams ES, Martin KA et al. Hip3 interacts with the HIRA proteins Hip1 and Slm9 and is required for transcriptional silencing and accurate chromosome segregation. J Biol Chem 2006; 281(13):8732-9.
80. Hou H, Wang Y, Kallgren SP et al. Histone variant H2A.Z regulates centromere silencing and chromosome segregation in fission yeast. J Biol Chem 285(3):1909-18.
81. Mehta M, Kim HS, Keogh MC. Sometimes one just isn't enough: do vertebrates contain an H2A.Z hyper-variant? J Biol 9(1):3.
82. Svotelis A, Gevry N, Gaudreau L. Regulation of gene expression and cellular proliferation by histone H2A.Z. Biochem Cell Biol 2009; 87(1):179-88.
83. Raisner RM, Hartley PD, Meneghini MD et al. Histone variant H2A.Z marks the 5′ ends of both active and inactive genes in euchromatin. Cell 2005; 123(2):233-48.
84. Pan X, Ye P, Yuan DS et al. A DNA integrity network in the yeast Saccharomyces cerevisiae. Cell 2006; 124(5):1069-81.
85. Reis CC, Campbell JL. Contribution of Trf4/5 and the nuclear exosome to genome stability through regulation of histone mRNA levels in Saccharomyces cerevisiae. Genetics 2007; 175(3):993-1010.
86. Lemay JF, D'Amours A, Lemieux C et al. The nuclear poly(A)-binding protein interacts with the exosome to promote synthesis of noncoding small nucleolar RNAs. Mol Cell 37(1):34-45.
87. Struhl K. Transcriptional noise and the fidelity of initiation by RNA polymerase II. Nat Struct Mol Biol 2007; 14(2):103-5.
88. Azzalin CM, Reichenbach P, Khoriauli L et al. Telomeric repeat containing RNA and RNA surveillance factors at mammalian chromosome ends. Science 2007; 318(5851):798-801.
89. Luke B, Panza A, Redon S et al. The Rat1p 5′ to 3′ exonuclease degrades telomeric repeat-containing RNA and promotes telomere elongation in Saccharomyces cerevisiae. Mol Cell 2008; 32(4):465-77.

90. Schoeftner S, Blasco MA. Chromatin regulation and noncoding RNAs at mammalian telomeres. Semin Cell Dev Biol 2009.
91. Schoeftner S, Blasco MA. A 'higher order' of telomere regulation: telomere heterochromatin and telomeric RNAs. Embo J 2009; 28(16):2323-36.
92. Schoeftner S, Blasco MA. Developmentally regulated transcription of mammalian telomeres by DNA-dependent RNA polymerase II. Nat Cell Biol 2008; 10(2):228-36.
93. Deng Z, Norseen J, Wiedmer A et al. TERRA RNA binding to TRF2 facilitates heterochromatin formation and ORC recruitment at telomeres. Mol Cell 2009; 35(4):403-13.
94. Garcia-Cao M, O'Sullivan R, Peters AH et al. Epigenetic regulation of telomere length in mammalian cells by the Suv39h1 and Suv39h2 histone methyltransferases. Nat Genet 2004; 36(1):94-9.
95. Singer MS, Gottschling DE. TLC1: template RNA component of Saccharomyces cerevisiae telomerase. Science 1994; 266(5184):404-9.
96. Singer MS, Kahana A, Wolf AJ et al. Identification of high-copy disruptors of telomeric silencing in Saccharomyces cerevisiae. Genetics 1998; 150(2):613-32.
97. Azzalin CM, Lingner J. The double life of UPF1 in RNA and DNA stability pathways. Cell Cycle 2006; 5(14):1496-8.
98. Azzalin CM, Lingner J. The human RNA surveillance factor UPF1 is required for S phase progression and genome stability. Curr Biol 2006; 16(4):433-9.
99. Redon S, Reichenbach P, Lingner J. Protein RNA and protein protein interactions mediate association of human EST1A/SMG6 with telomerase. Nucleic Acids Res 2007; 35(20):7011-22.
100. Berretta J, Morillon A. Pervasive transcription constitutes a new level of eukaryotic genome regulation. EMBO Rep 2009; 10(9):973-82.
101. San Paolo S, Vanacova S, Schenk L et al. Distinct roles of noncanonical poly(A) polymerases in RNA metabolism. PLoS Genet 2009; 5(7):e1000555.
102. Chapon C, Cech TR, Zaug AJ. Polyadenylation of telomerase RNA in budding yeast. RNA 1997; 3(11):1337-51.
103. Askree SH, Yehuda T, Smolikov S et al. A genome-wide screen for Saccharomyces cerevisiae deletion mutants that affect telomere length. Proc Natl Acad Sci USA 2004; 101(23):8658-63.
104. Berretta J, Pinskaya M, Morillon A. A cryptic unstable transcript mediates transcriptional trans-silencing of the Ty1 retrotransposon in S. cerevisiae. Genes Dev 2008; 22(5):615-26.
105. Camblong J, Beyrouthy N, Guffanti E et al. Trans-acting antisense RNAs mediate transcriptional gene cosuppression in S. cerevisiae. Genes Dev 2009; 23(13):1534-45.
106. Camblong J, Iglesias N, Fickentscher C et al. Antisense RNA stabilization induces transcriptional gene silencing via histone deacetylation in S. cerevisiae. Cell 2007; 131(4):706-17.
107. Sun BK, Deaton AM, Lee JT. A transient heterochromatic state in Xist preempts X inactivation choice without RNA stabilization. Mol Cell 2006; 21(5):617-28.
108. Ng K, Pullirsch D, Leeb M et al. Xist and the order of silencing. EMBO Rep 2007; 8(1):34-9.
109. de Napoles M, Mermoud JE, Wakao R et al. Polycomb group proteins Ring1A/B link ubiquitylation of histone H2A to heritable gene silencing and X inactivation. Dev Cell 2004; 7(5):663-76.
110. Marks H, Chow JC, Denissov S et al. High-resolution analysis of epigenetic changes associated with X inactivation. Genome Res 2009; 19(8):1361-73.
111. Ciaudo C, Bourdet A, Cohen-Tannoudji M et al. Nuclear mRNA degradation pathway(s) are implicated in Xist regulation and X chromosome inactivation. PLoS Genet 2006; 2(6):e94.
112. Morey C, Arnaud D, Avner P et al. Tsix-mediated repression of Xist accumulation is not sufficient for normal random X inactivation. Hum Mol Genet 2001; 10(13):1403-11.
113. Ogawa Y, Sun BK, Lee JT. Intersection of the RNA interference and X-inactivation pathways. Science 2008; 320(5881):1336-41.
114. Nesterova TB, Popova BC, Cobb BS et al. Dicer regulates Xist promoter methylation in ES cells indirectly through transcriptional control of Dnmt3a. Epigenetics Chromatin 2008; 1(1):2.
115. Ilik I, Akhtar A. roX RNAs: noncoding regulators of the male X chromosome in flies. RNA Biol 2009; 6(2):113-21.
116. Mendjan S, Taipale M, Kind J et al. Nuclear pore components are involved in the transcriptional regulation of dosage compensation in Drosophila. Mol Cell 2006; 21(6):811-23.
117. Iida T, Kawaguchi R, Nakayama J. Conserved ribonuclease, Eri1, negatively regulates heterochromatin assembly in fission yeast. Curr Biol 2006; 16(14):1459-64.
118. Ibrahim F, Rymarquis LA, Kim EJ et al. Uridylation of mature miRNAs and siRNAs by the MUT68 nucleotidyltransferase promotes their degradation in Chlamydomonas. Proc Natl Acad Sci USA.
119. van Wolfswinkel JC, Claycomb JM, Batista PJ et al. CDE-1 affects chromosome segregation through uridylation of CSR-1-bound siRNAs. Cell 2009; 139(1):135-48.

CHAPTER 10

CONTROL OF CRYPTIC TRANSCRIPTION IN EUKARYOTES

Mathieu Rougemaille* and Domenico Libri*

LEA Laboratory of Nuclear RNA Metabolism, Centre de Génétique Moléculaire, CNRS-UPR2167, Gif-sur-Yvette, France; and Centre for mRNP Biogenesis and Metabolism, Department of Molecular Biology, Aarhus University, Aarhus, Denmark.

**Correspondence—Emails: rougemai@cgm.cnrs-gif.fr, libri@cgm.cnrs-gif.fr.*

Abstract: Over the last few years, the development of large-scale technologies has radically modified our conception of genome-wide transcriptional control by unveiling an unexpected high complexity of the eukaryotic transcriptome. In organisms ranging from yeast to human, a considerable number of novel small RNA species have been discovered in regions that were previously thought to be incompatible with high levels of transcription. Intriguingly, these transcripts, which are rapidly targeted for degradation by the exosome, appear to be devoid of any coding potential and may be the consequence of unwanted transcription events. However, the notion that an important fraction of these RNAs represent by-products of regulatory transcription is progressively emerging. In this chapter, we discuss the recent advances made in our understanding of the shape of the eukaryotic transcriptome. We also focus on the molecular mechanisms that cells exploit to prevent cryptic transcripts from interfering with the expression of protein-coding genes. Finally, we summarize data obtained in different systems suggesting that such RNAs may play a critical role in the regulation of gene expression as well as the evolution of genomes.

INTRODUCTION: NEW PERSPECTIVES ON THE SHAPE OF THE TRANSCRIPTOME

The genetic information encoded in the DNA has been classically defined by two major parameters: first, the "punctuation" of the DNA string (e.g., transcription initiation and termination sites and the position and nature of processing sites), which defines the fraction of the genome encoding informative RNAs; and second, the expression level

RNA Exosome, edited by Torben Heick Jensen.

of the gene and its regulation, which has traditionally been linked to transcription rates. The assessment of these qualitative and quantitative parameters has mostly relied on the product of transcription, i.e., the sequence and steady state abundance of the mature RNA and its function as a coding or noncoding molecule.

This conception has been partially challenged by the recent discovery of "hidden" transcription in the genome.[1-9] It is becoming increasingly clear that the fraction of the genome that is transcribed is much larger than previously thought in organisms ranging from yeast to human. Classical punctuation rules (i.e., the definition of the "borders" of a gene) are also challenged by the discovery of widespread overlapping and antisense transcription. The establishment of this new concept is historically connected to conceptual and technological advances in two important parameters of analysis. First, considering the stability of transcripts, most often by analyzing the transcriptome in mutants of the RNA degradation machinery; i.e., the exosome. In fact, the absence of a given RNA species does not necessarily mean that the transcript is not produced (and therefore that there is no gene coding for it), but could also imply that the RNA is transcribed and degraded so rapidly and efficiently that it cannot be detected using conventional biochemical methods. The second—and related—parameter is the possibility to directly assess the position of the transcription machinery by a plethora of approaches including chromatin immunoprecipitation (ChIP), genomic nuclear run on (NRO) and large scale sequencing analyses. Therefore, inferring the presence of RNA polymerase (RNAP) in a given region of the genome is not anymore uniquely defined by the detection of an associated and stable RNA molecule. Importantly, these new approaches are open ended, as they are generally not biased by the preconception of transcription occurrence. Thus, RNAPs and/or RNA molecules have been frequently detected in regions of the genome and in orientations that where previously not suspected or even thought to be incompatible with efficient gene expression, e.g., antisense to known ORFs or in heterochromatin regions. This almost ubiquitous presence of RNAPs, defined "pervasive", has been suggested to result from a leaky control on transcription initiation, or from the inherent ability of polymerases to transcribe in both directions once the assembly of the transcription machinery on a given promoter has taken place[4,6-9] (see below).

A DEDICATED PATHWAY OF NUCLEAR RNA DEGRADATION

RNA decay is an important factor defining the steady state abundance of cellular transcripts. Efficient RNA degradation is also necessary to allow rapid changes in transcript amount and failure to control RNA stability has been implicated in many diseases, including cancer (for reviews see refs. 10,11). In metazoans, most RNA degradation occurs in the nucleus (for review see ref. 12) as illustrated by all the products of processing that have to be discarded: introns, endonucleolytic fragments and portions of primary transcripts that are removed exonucleolytically. Nuclear degradation of RNA also occurs in yeast, but perhaps less so than in metazoans as noncoding segments (essentially introns) are generally less extensive than coding regions. However, even though the recent discovery of cryptic transcription in yeast has significantly altered this balance, a more precise estimate must await quantitative analyses on transcription levels of stable and unstable transcripts.

The main actor in nuclear RNA decay is the exosome, a conserved complex comprising a dozen of subunits that has been implicated in a large number of RNA degradation and processing pathways (for recent reviews see refs. 13,14 as well as other chapters in this book).

Present in both the nucleus and the cytoplasm,[15] the exosome complex consists of a core containing ten polypeptides, but only the Dis3p/Rrp44p subunit provides a processive hydrolytic 3′ to 5′ exoribonuclease activity.[16-18] Importantly, structural and biochemical analyses suggested that RNAs targeted for degradation thread through the central channel of the core to reach the Dis3p exoribonuclease site.[18] In addition, the RNase D homologue Rrp6p (PM/Scl-100 in humans), the eleventh subunit of the nuclear exosome,[19] contributes a distributive hydrolytic exoribonuclease activity.[16,17,20] However, the catalytic activity of the exosome is not uniquely related to 3′-5′ decay as recent studies have led to the discovery that Dis3p is also endowed with an endoribonuclease activity mediated by a conserved PIN domain.[21-23] Both endo- and exonuclease activities of Dis3p appear to actively cooperate in exosome functions as yeast strains carrying mutations in the two catalytic domains are nonviable. As expected, processing of various rRNAs substrates as well as degradation of products resulting from cryptic transcription are defective in those mutants[21-23] (see below). The relative contribution of these two enzymatic activities in vivo is still poorly understood: it is possible that the different activities of the Dis3p subunit are responsible for degradation of different classes of RNA substrates. Alternatively, the endonuclease activity might serve as a backup mechanism that allows degradation of RNA species that contain highly structured regions impeding the progression of 3′ to 5′ decay.

The wide variety of transcripts undergoing exosome-mediated decay suggests that several distinct pathways detect and target substrates to the exosome complex. Supporting such versatility, a number of different exosome cofactors have been identified during the last decade. Among these is the conserved TRAMP (Trf4p Air1-2p Mtr4p Polyadenylation) complex that has been functionally linked to the nuclear exosome.[2,24,25] This complex is composed of one poly(A) polymerase, Trf4p or Trf5p, the DEVH-box ATP-dependent putative RNA helicase Mtr4p and one of the putative RNA binding proteins Air1p or Air2p. Akin to the poly(A)-dependent degradation of mRNAs in *E. coli*,[26] it is believed that the TRAMP complex stimulates exosome activity by adding a poly(A) tail on substrates targeted for degradation, thus facilitating their remodeling and the progression of the exosome through regions of secondary structure.[24] TRAMP has been involved in the polyadenylation-assisted degradation of various RNA species including hypomodified tRNAs,[27,28] precursors of snoRNAs, snRNAs and rRNAs,[2,24,25,28] as well as the so-called CUTs (Cryptic Unstable Transcripts) which are very short-lived RNAPII products that were originally identified in several intergenic regions[2] (see below). Interestingly, the stimulatory activity of TRAMP on exosome function is not only linked to its polyadenylation function, but might also relate to addressing the degradation enzyme to its substrates, possibly through the RNA-binding activity of the Air1/2p proteins.[29]

WIDESPREAD TRANSCRIPTION IN EUKARYOTIC GENOMES

The concept of hidden transcription, i.e., the existence of short-lived RNA molecules that are not detectable at steady state in a wild type environment, has emerged early during the last decade based on the observation of large-scale transcriptional activity in human cells.[1] Evidence for hidden transcription in yeast first came from microarray analysis of the transcriptome of strains lacking the Rrp6p nuclear subunit of the exosome. Increased expression of numerous transcripts arising from intergenic regions, previously thought to be silent, was detected in this mutant.[2] Importantly, similar levels of RNAPII were observed by ChIP in the *rrp6Δ* strain when compared to a wild type strain, suggesting that detection of these RNAs does not result from transcriptional up-regulation.[2] These findings led to the surprising conclusion that some transcripts are produced and rapidly targeted for degradation following transcription. For these reasons, they were dubbed CUTs. Subsequent characterization established that CUTs have extremely short half-lives, are relatively small in size (200-600 nucleotides), possess a cap at their 5′ end and have heterogenous 3′ ends. Importantly, these 3′ ends are polyadenylated in a Trf4p-dependent manner and, consistent with this observation, degradation of CUTs depends on the combined action of the TRAMP and exosome complexes.[2] More recent work has extended these findings by showing that the Dis3-dependent endonucleolytic activity of the exosome is also required for efficient degradation of these enigmatic transcripts.[21]

Five years after their discovery, it is now known that the occurrence of CUTs is widespread in the yeast genome. Originally estimated to derive from at least 10% of all intergenic regions,[2] CUTs were subsequently shown to be present in other regions of the genome: promoter-associated, overlapping or antisense to known ORFs.[3,30-32] Recent work performed in the Steinmetz and Jacquier laboratories,[8,9] together with our unpublished observations, has allowed the definition of a complete map of the distribution of CUTs throughout the genome. This has led to the surprising conclusion that CUTs are as abundant as other RNAPII transcripts, indicating that a large share of polymerase activity is devoted to the production of RNAs with no coding potential and possibly no function. Another important conclusion from these studies is that most promoters in yeast are bi-directional, expressing, depending on the sequence context, any possible combinations of stable or unstable transcripts.[8,9] The pattern of CUTs distribution highlights the potential role of these elements as regulators but also as products of unwanted transcription initiation events.

The observation of abundant transcription from intergenic regions of mammalian chromosomes (for review see ref. 33) first suggested the evolutionary conservation of widespread transcription. Recent landmark genome-wide studies have confirmed the generality of this phenomenon and revealed a previously unanticipated level of complexity of the metazoan transcriptome.[4-7] A considerable fraction of antisense and promoter-associated transcripts have indeed been detected in various organisms, including flies, mice and humans.[4-7,34]

In analogy to the yeast model, human cells depleted for the hRrp40p core subunit of the exosome accumulate polyadenylated unstable transcripts called PROMPTs (PROMoter uPstream Transcripts) arising from both strands, upstream of many, if not all active, gene promoters.[6] Although the sizes of these RNAs are so far unknown, the regions from which PROMPTs derive are characterized by chromatin modifications associated with transcription initiation but not elongation, suggesting that they are relatively small. The physiological relevance of PROMPTs remains elusive but it is possible that, like CUTs in yeast, they represent products of spurious transcription events or somehow regulatory elements of gene expression.

TRANSCRIPTION TERMINATION AND THE HALLMARK OF CUTs INSTABILITY

One important question regarding CUTs instability is how they are distinguished from stable noncoding RNAs and mRNAs. What are the signals and determinants responsible for their rapid degradation by the TRAMP and exosome complexes? Several studies have shed light on the general mechanism that leads to the degradation of CUTs and to their recognition as unstable RNAs. Indeed, it was shown that the extreme instability of CUTs is directly linked to the mechanism of transcription termination of the loci encoding these RNAs.[31,36] In yeast, there are two major pathways that elicit termination of transcription: one that is dedicated to mRNA production and requires the cleavage and polyadenylation complex (CPF/CF) and one that directs transcription termination of stable noncoding RNAs, such as sno- and sn-RNAs.[37] This latter pathway involves the action of the Nrd1p complex that contains two RNA binding proteins, Nrd1p and Nab3p and one helicase, Sen1p.[37,38] Nrd1p interacts directly with the RNAPII carboxy terminal domain (CTD)[39,40] and recognizes GUAA/G sequences on the RNA, while Nab3p binds to UCUU motifs.[38] It has been proposed that binding to both the nascent transcript and the polymerase leads to the release of the transcription machinery from the DNA template. In addition, the Nrd1p-Nab3p-Sen1p complex also associates with the nuclear exosome,[41] which is thought to couple transcription termination to 3′-5′ exonucleolytic degradation/processing. In the case of snoRNAs, it is believed that the assembly of factors onto the nascent RNA molecule impedes the progression of the exosome thereby defining the 3′ end of the mature RNA.

Transcription termination of CUTs also requires the Nrd1-Nab3-dependent pathway.[31,36] CUTs contain multiple Nrd1p and Nab3p binding sites in their sequences and mutation of these motifs leads to both transcription termination failure and stabilization of longer species deriving from downstream Nrd1p-independent transcription termination. Further supporting a direct role for this termination pathway, the absence of a functional Nrd1p complex also strongly stabilizes CUTs as longer RNA species.[31,36] Thus, a pathway that directs 3′-end processing of stable noncoding RNAs (sn- and snoRNAs) is also devoted to control pervasive transcription, which, considering the large number of transcription units producing CUTs, might indeed be its primary function. Interestingly, transcription arising spuriously within genes, as in mutants having a defective chromatin structure (e.g., the histone chaperones Spt6p and Spt16p),[42,43] does not produce unstable RNAs. This is presumably because in these cases the cryptic transcription events use the 3'-end termination signals of the host genes, underscoring the linkage between stability and 3′-end processing.

Since degradation is coupled to transcription termination, the RNAs produced are degraded extremely fast after their production, ensuring that they interact for only a short period of time with the gene expression pathways of the cell. This fits with both hypotheses that these RNAs are discard products of spurious transcription events, or short-lived regulatory molecules (see below).

AN RNAPII "CTD CODE" FOR THE FATE OF NASCENT TRANSCRIPTS

Because Nrd1p-Nab3p binding sites are very frequent in the yeast genome, the robust expression of coding genes that are often several kilobases long is potentially jeopardized by their presence. Which mechanisms have cells implemented to avoid the spurious occurrence of Nrd1p-dependent transcription termination events? One possibility would be to select against Nrd1p and Nab3p binding sites in coding regions, but this would imply a high evolutionary cost for long genes. Rather, recent evidence indicates that Nrd1p-dependent transcription termination depends on a mechanism that senses the length of the transcribed gene and confers immunity to Nrd1p-dependent termination (and nuclear degradation) once a given threshold has been reached[44] (Fig. 1). Biochemical and genetic analyses have revealed that such a mechanism is based on

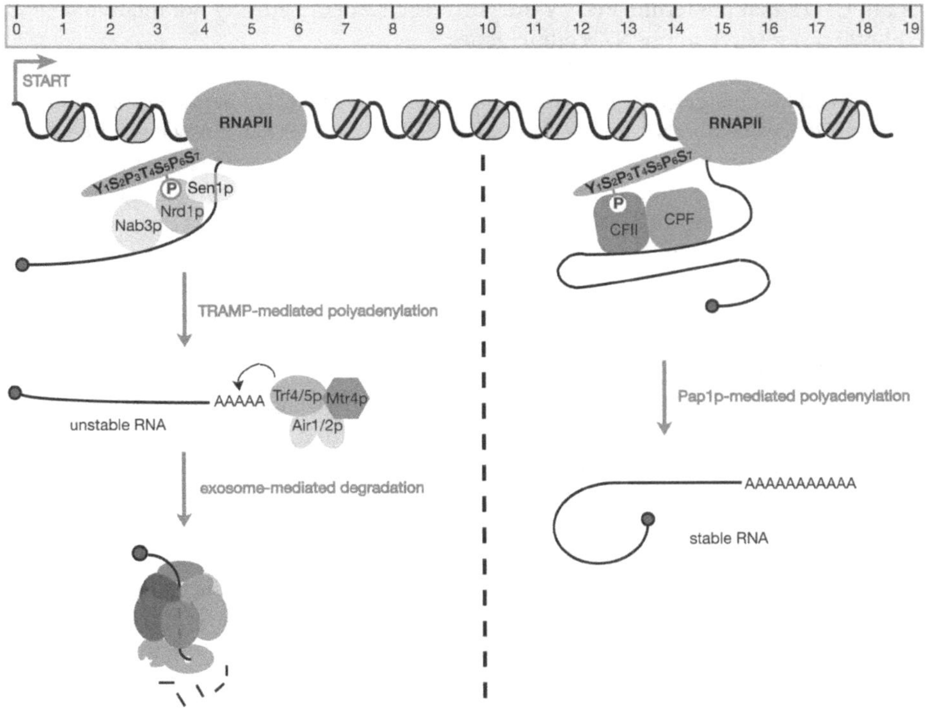

Figure 1. Schematic representation of the fate of nascent transcripts according to the transcription termination pathway they are targeted to. Two complexes mediate transcription termination of nascent RNA molecules: the Nrd1p-Nab3p-Sen1p complex that associates with RNAPII during the early phases of transcription when the CTD is phosphorylated on Ser 5 (left) and the CF/CPF complex that interacts with the CTD when it is phosphorylated on Ser 2, during the late phases of the transcription cycle (right). In the case of CUTs, Nrd1p-dependent termination occurs soon after transcription elongation has started and promotes the release of the transcript that is subsequently targeted for TRAMP-dependent polyadenylation and exosome-mediated degradation (left). However, once a given threshold has been reached, RNAPII becomes insensitive to Nrd1p termination signals and the nascent RNA is subjected to termination by the CF/CPF complex. This is associated with a polyadenylation step mediated by the canonical poly(A) polymerase Pap1p that promotes stabilization of the transcript (right). The ruler at the top of the schema illustrates this "length effect" that governs the fate of nascent RNA molecules.

the phosphorylation status of the RNAPII CTD. This domain is composed of several repeats (26 in yeast and 52 in mammals) of a heptapeptide (YSPTSPS) that contains serine (Ser) residues at position two, five and seven, each of which are substrates for different kinases acting at various stages of the transcription cycle. During the early phases of transcription, the RNAPII CTD is mainly phosphorylated at Ser 5, while late steps in transcription are associated with Ser 2 phosphorylation. It has been shown that termination by the Nrd1p complex is preferentially occurring during the early phase of transcription, when the CTD is phosphorylated on Ser 5.[40,44] Hence, the polymerase becomes insensitive to the presence of Nrd1p and Nab3p binding sites during the late phases of transcription, when phosphorylation is mainly directed to Ser 2.[44] Therefore, an "odometer" for transcription is provided by the phosphorylation status of the CTD: sensing the position of RNAPII determines sensitivity to Nrd1p-dependent termination until a given threshold of transcription (Fig. 1). Consistent with this finding, most, if not all, CUTs that are terminated by the Nrd1p-dependent pathway are relatively short (200-600 bp) relative to mRNA coding genes.

In addition to phosphorylation of Ser residues, the RNAPII CTD can also be targeted for noncovalent changes, such as prolyl isomerization. The CTD heptapeptide contains two Ser-Pro peptide bonds (i.e., S_2P_3 and S_5P_6), which are potential substrates for peptidyl prolyl *cis/trans* isomerases. This modification is believed to trigger conformational changes within the CTD that may affect the binding of RNAPII cofactors (for review see ref. 45). In budding yeast, the essential CTD prolyl isomerase Ess1p preferentially binds and isomerizes phospho-S_5P_6 within the CTD, when compared to phospho-S_2P_3.[46] Interestingly, recent work demonstrated that Ess1p is required for Nrd1p-dependent termination of noncoding RNAs, including snoRNAs and CUTs, presumably by controlling the phosphorylation status of the CTD Ser 5.[47,48]

Together, these findings have important consequences on the mechanisms by which the cell control the occurrence of spurious (or regulatory) transcription while permitting the robust expression of "established" genetic information. The Nrd1p complex only impacts transcription events for which the otherwise frequent termination signals have not been counter-selected in the early region of the transcribed region. However, after transcription has proceeded beyond a given threshold, termination cannot occur by this pathway and the production of a stable transcript is ensured (Fig. 1). In light of the confirmed existence of pervasive transcription, it is particularly significant that extremely sensitive mechanisms exist to cope with "lost" polymerases (transcription termination) and potentially toxic transcripts (degradation). The "length effect", intimately associated with a "CTD code", provides a fail-safe mechanism to protect protein-coding genes from such a quality control.

REGULATORY ROLES OF CRYPTIC TRANSCRIPTION

The biological significance of cryptic transcription is still the subject of debate. Does it reflect background transcriptional noise or is it the source of novel regulatory pathways for the control of gene expression? While a number of cryptic RNAs and transcription events are likely to derive from a leaky control on transcription repression by chromatin structure, it is becoming increasingly clear that at least some unstable noncoding transcripts are implicated in the regulation of gene expression. Although there is little evidence that the most abundant class of CUTs, that is the one produced by divergent transcription, is functional, the less frequent sense CUTs have in some cases been shown to associate with regulatory events. For instance, it was shown that expression of the *SRG1* locus (producing noncoding RNAs some of which are CUTs) prevents transcriptional activation of the downstream *SER3* gene by blocking the binding of transcription factors and pre-initiation complex formation (PIC) at the promoter.[49,50] This mechanism of gene regulation constituted the first example of transcription interference in budding yeast and recent large-scale analyses have indeed revealed that the expression of sense CUTs and their associated mRNAs are often anti-correlated.[8]

Another mechanism of regulation that implicates CUTs occurs at several genes belonging to the nucleoside triphosphate biosynthesis pathway, including *IMD2*, *URA2*, *URA8* and *ADE8* . In these cases, the expression of a sense CUT and the downstream mRNA depends on the same promoter and PIC, but surprisingly, is driven by distinct transcription start sites (TSS).[32,51,52] Upon nucleotide starvation, transcription initiates at a downstream site and a functional mRNA is produced. However, in conditions of high nucleotide concentrations, transcription initiation occurs at upstream sites and leads to the accumulation of unstable transcripts (CUTs) that are not protein encoding. Nrd1p and Nab3p-binding sites between the two alternative TSSs ensure early transcription termination and rapid degradation of the CUT, but are skipped when the downstream start site is selected, allowing the production of a long and functional mRNA. Thus, contrary to most known cases in which regulation of gene expression occurs at the level of promoter activation, transcription of genes involved in nucleotide biogenesis is constitutively activated but is diverted on a nonproductive pathway in repressing conditions. The event that leads to the functional activation of the gene is a switch in the selection of the transcription start site, which is unprecedented in eukaryotes.

In addition to the mechanisms listed above, the production of antisense RNAs is also a source of gene regulation. Indeed, antisense transcription at the *IME4* and *PHO84* loci[53,54] has been shown to regulate expression of the cognate sense transcripts, either by transcriptional interference (*IME4*) or by an epigenetic mechanism that involves the recruitment of chromatin-modifying enzymes to prevent transcriptional activation (*PHO84*). Only in the latter example, the noncoding RNA plays an essential role in regulation by recruiting a histone de-acetylase. At the *IME4* locus, regulation is dependent on antisense transcription more than on the transcript that can be considered as a by-product.

Although most studies on the role of cryptic transcripts have been performed in yeast, specific examples of regulatory intergenic transcription in *Drosophila*, mouse and humans have also been described.[55-57] Moreover, specific human promoter-associated transcripts have been shown to downregulate the expression of adjacent genes.[58] Therefore, pervasive transcription might play important regulatory roles in gene expression throughout the eukaryotic kingdom.

CONCLUSION

The classical view of the expression of genetic information has been profoundly challenged by the recent discovery of widespread transcription and the identification of numerous novel RNA species. Hidden and pervasive transcription are now known to be a common feature of eukaryotic transcriptomes.[2,4-7,35] These discoveries have resulted in a major upheaval in the field and led to the reconsideration of the basic mechanisms of transcription and its regulation. However, so far the findings raise more fundamental questions than they give answers. How can the cell cope with a multitude of converging polymerases that produce overlapping or complementary transcripts? How are potential collisions between converging transcription machineries dealt with? Is the potential interference of nearby expression units exploited to evolve complex regulatory pathways? Why does the cell "correct" spurious RNAP transcription initiation rather than implement an error-proof mechanism? An intriguing hypothesis is that error-prone mechanisms of transcription initiation generating cryptic transcripts are maintained both because of their more "economical" nature and because this might allow more opportunities for evolution. Transcribing more than the "established" genes provides a chance to expose additional, potentially useful information to selective pressure, possibly providing an opportunity for generating new functions. Accordingly, the combination of transcription termination and degradation mechanisms ensures the elimination of unwanted RNA molecules without precluding the process of transcription itself. In this perspective, cryptic transcripts can be considered as potential "genes-to-be" to the same extent as random mutations generate genetic variability that will be subjected to the selective pressure. A major challenge for future studies will be to understand whether cryptic transcription widely contributes to the regulation of gene expression and whether it impacts the evolutionary process that may allow the acquisition of novel functions through yet unidentified mechanisms.

REFERENCES

1. Kapranov P, Cawley SE, Drenkow J et al. Large-scale transcriptional activity in chromosomes 21 and 22. Science 2002; 296:916-919.
2. Wyers F, Rougemaille M, Badis G et al. Cryptic pol II transcripts are degraded by a nuclear quality control pathway involving a new poly(A) polymerase. Cell 2005; 121:725-737.
3. Davis CA, Ares M Jr. Accumulation of unstable promoter-associated transcripts upon loss of the nuclear exosome subunit Rrp6p in Saccharomyces cerevisiae. Proc Natl Acad Sci USA 2006; 103:3262-3267.
4. Core LJ, Waterfall JJ, Lis JT. Nascent RNA sequencing reveals widespread pausing and divergent initiation at human promoters. Science 2008; 322:1845-1848.
5. He Y, Vogelstein B, Velculescu VE et al. The antisense transcriptomes of human cells. Science 2008; 322:1855-1857.
6. Preker P, Nielsen J, Kammler S et al. RNA exosome depletion reveals transcription upstream of active human promoters. Science 2008; 322:1851-1854.
7 Seila AC, Calabrese JM, Levine SS et al. Divergent transcription from active promoters. Science 2008; 322:1849-1851.
8 Neil H, Malabat C, d'Aubenton-Carafa Y et al. Widespread bidirectional promoters are the major source of cryptic transcripts in yeast. Nature 2009; 457:1038-1042.
9. Xu Z, Wei W, Gagneur J et al. Bidirectional promoters generate pervasive transcription in yeast. Nature 2009; 457:1033-1037.
10. Hollams EM, Giles KM, Thomson AM et al. mRNA stability and the control of gene expression: implications for human disease. Neurochem Res 2002; 27:957-980.
11. Audic Y, Hartley RS. Post-transcriptional regulation in cancer. Biol Cell 2004; 96:479-498.
12. Moore MJ. Nuclear RNA turnover. Cell 2002; 108:431-434.
13. Schmitt M, Jensen TH. The exosome: a multipurpose RNA-decay machine. Trens Biochem Sci 2008; 33:501-510.
14. Lebreton A, Séraphin B. Exosome-mediated quality control: substrate recruitment and molecular activity. Biochim Biophys Acta 2008; 1179:558-565.
15. Allmang C, Petfalski E, Podtelejnikov A et al. The yeast exosome and human PM-Scl are related complexes of 3′ —> 5′ exonucleases. Genes Dev 1999; 13:2148-2158.
16. Liu Q, Greimann JC, Lima CD. Reconstitution, activities and structure of the eukaryotic RNA exosome. Cell 2006; 127:1223-1237.
17. Dziembowski A, Lorentzen E, Conti E et al. A single subunit, Dis3, is essentially responsible for yeast exosome core activity. Nat Struct Mol Biol 2007; 14:15-22.
18. Bonneau F, Basquin J, Ebert J et al. The yeast exosome functions as a macromolecular cage to channel RNA substrates for degradation. Cell 2009; 139:547-559.
19. Mitchell P, Tollervey D. Musing on the structural organization of the exosome complex. Nat Struct Biol 2000; 7:843-846.
20. Burkard KT, Butler JS. A nuclear 3′-5′ exonuclease involved in mRNA degradation interacts with Poly(A) polymerase and the hnRNA protein Npl3p. Mol Cell Biol 2000; 20:604-616.
21. Lebreton A, Séraphin B. Endonucleolytic RNA cleavage by a eukaryotic exosome. 2008; 456:993-996.
22. Schaeffer D, Tsanova B, Barbas A et al. The exosome contains domains with specific endoribonuclease, exoribonuclease and cytoplasmic mRNA decay activities. Nat Struct Mol Biol 2008; 16:56-62.
23. Schneider C, Leung E, Brown J et al. The N-terminal PIN domain of the exosome subunit Rrp44 harbors endonuclease activity and tethers Rrp44 to the yeast core exosome. Nucleic Acids Res 2009; 37:1127-1140.
24. LaCava J, Houseley J, Saveanu C et al. RNA degradation by the exosome is promoted by a nuclear polyadenylation complex. Cell 2005; 121:713-724.
25. Vanacova S, Wolf J, Martin G et al. A new yeast poly(A) polymerase complex involved in RNA quality control. PLoS Biol 2005; 3:e189.
26. Carpousis AJ, Vanzo NF, Raynal LC. mRNA degradation. A tale of poly(A) and multiprotein machines. Trends Genet 1999; 15:24-28.
27. Kadaba S, Krueger A, Trice T et al. Nuclear surveillance and degradation of hypomodified initiator tRNAMet in S. cerevisiae. Genes Dev 2004; 18:1227-1240.

28. Kadaba S, Wang X, Anderson J. Nuclear RNA surveillance in Saccharomyces cerevisiae: Trf4p-dependent polyadenylation of nascent hypomethylated tRNA and an aberrant form of 5S rRNA. RNA 2006; 12:508-521.
29. Rougemaille M, Gudipati RK, Olesen JR et al. Dissecting mechanisms of nuclear mRNA surveillance in THO/sub2 complex mutants. Embo J 2007; 26:2317-2326.
30. Houalla R, Devaux F, Fatica A et al. Microarray detection of novel nuclear RNA substrates for the exosome. Yeast 2006; 23:439-454.
31. Arigo JT, Eyler DE, Carroll KL et al. Termination of cryptic unstable transcripts is directed by yeast RNA-binding proteins Nrd1 and Nab3. Mol Cell 2006; 23:841-851.
32. Thiebaut M, Colin J, Neil H et al. Futile cycle of transcription initiation and termination modulates the response to nucleotide shortage in S. cerevisiae. Mol Cell 2008; 31:671-682.
33. Johnson JM, Edwards S, Shoemaker D et al. Dark matter in the genome: evidence of widespread transcription detected by microarray tiling experiments. Trends Genet 2005; 21:93-102.
34. Taft RJ, Glazov EA, Cloonan N, Simons C et al. Tiny RNAs associated with transcription start sites in animals. Nat Genet 2009; 41:572-578.
35. Kapranov P, Cheng J, Dike S et al. RNA maps reveal new RNA classes and a possible function for pervasive transcription. Science 2007; 316:1484-1488.
36. Thiebaut M, Kisseleva-Romanova E, Rougemaille M et al. Transcription termination and nuclear degradation of cryptic unstable transcripts: a role for the nrd1-nab3 pathway in genome surveillance. Mol Cell 2006; 23:853-864.
37. Steinmetz EJ, Conrad NK, Brow DA et al. RNA-binding protein Nrd1 directs poly(A)-independent 3′-end formation of RNA polymerase II transcripts. Nature 2001; 413:327-331.
38. Carroll KL, Pradhan DA, Granek JA et al. Identification of cis elements directing termination of yeast nonpolyadenylated snoRNA transcripts. Mol Cell Biol 2004; 24:6241-6252.
39. Yuryev A, Patturajan M, Litingtung Y et al. The C-terminal domain of the largest subunit of RNA polymerase II interacts with a novel set of serine/arginine-rich proteins. Proc Natl Acad Sci USA 1996; 93:6975-6980.
40. Vasiljeva L, Kim M, Mutschler H et al. The Nrd1-Nab3-Sen1 termination complex interacts with the Ser5-phosphorylated RNA polymerase II C-terminal domain. Nat Struct Mol Biol 2008; 15:795-804.
41. Vasiljeva L, Buratowski S. Nrd1 interacts with the nuclear exosome for 3′ processing of RNA polymerase II transcripts. Mol Cell 2006; 21:239-248.
42. Kaplan CD, Laprade L, Winston F. Transcription elongation factors repress transcription initiation from cryptic sites. Science 2003; 301:1096-1099.
43. Mason PB, Struhl K. The FACT complex travels with elongating RNA polymerase II and is important for the fidelity of transcriptional initiation in vivo. Mol Cell Biol 2003; 23:8323-8333.
44. Gudipati RK, Villa T, Boulay et al. Phosphorylation of the RNA polymerase II C-terminal domain dictates transcription termination choice. Nat Struct Mol Biol 2008; 15:786-794.
45. Buratowski S. The CTD code. Nat Struct Biol 2003; 10:679-680.
46. Gemmill TR, Wu X, Hanes SD. Vanishingly low levels of Ess1 prolyl-isomerase activity are sufficient for growth in Saccharomyces cerevisiae. J Biol Chem 2005; 280:15510-15517.
47. Krishnamurthy S, Ghazy MA, Moore C et al. Functional interaction of the Ess1 prolyl isomerase with components of the RNA polymerase II initiation and termination machineries. Mol Cell Biol 2009; 29:2925-2934.
48. Singh N, Ma Z, Gemmill T et al. The Ess1 prolyl isomerase is required for transcription termination of small noncoding RNAs via the Nrd1 pathway. Mol Cell 2009; 36:255-266.
49. Martens JA, Laprade L, Winston F. Intergenic transcription is required to repress the Saccharomyces cerevisiae SER3 gene. Nature 2004; 429:571-574.
50. Martens JA, Wu PY, Winston F. Regulation of an intergenic transcript controls adjacent gene transcription in Saccharomyces cerevisiae. Genes Dev 2005; 19:2695-2704.
51. Jenks MH, O'Rourke TW, Reines D. Properties of an intergenic terminator and start site switch that regulate IMD2 transcription in yeast. Mol Cell Biol 2008; 28:3883-3893.
52. Kuehner JN, Brow DA. Regulation of a eukaryotic gene by GTP-dependent start site selection and transcription attenuation. Mol Cell 2008; 31:201-211.
53. Hongay CF, Grisafi PL, Galitski T et al. Antisense transcription controls cell fate in Saccharomyces cerevisiae. Cell 2006; 127:735-745.
54. Camblong J, Iglesias N, Fickentscher C et al. Antisense RNA stabilization induces transcriptional gene silencing via histone deacetylation in S. cerevisiae. Cell 2007; 131:706-717.
55. Gribnau J, Diderich K, Pruzina S et al. Intergenic transcription and developmental remodeling of chromatin subdomains in the human beta-globin locus. Mol Cell 2000; 5:377-386.
56. Bolland DJ, Wood AL, Johnston CM et al. Antisense intergenic transcription in V(D) J recombination. Nat Immunol 2004; 5:630-637.
57. Schmitt S, Prestel M, Paro R. Intergenic transcription through a polycomb group response element counteracts silencing. Genes Dev 2005; 19:697-708.
58. Wang X, Arai S, Song X et al. Induced ncRNAs allosterically modify RNA-binding proteins in cis to inhibit transcription. Nature 2008; 454:126-130.

CHAPTER 11

THE HUMAN EXOSOME AND DISEASE

Raymond H.J. Staals and Ger J.M. Pruijn*

Department of Biomolecular Chemistry, Nijmegen Center for Molecular Life Sciences, Institute for Molecules and Materials, Radboud University Nijmegen, Nijmegen, The Netherlands.
**Corresponding Author: Ger J.M. Pruijn—Email: g.pruijn@ncmls.ru.nl*

Abstract: Long before the RNA degrading exosome was first described in the yeast *Saccharomyces cerevisiae*, the use of autoantibodies found in the sera of certain autoimmune patients allowed the identification of a complex of polypeptides which later appeared to be the human exosome. Today, the most extensively documented association of the exosome with disease is still its targeting by the immune system of such patients. The highest frequency of autoantibodies to components of the exosome complex is found in polymyositis-scleroderma overlap patients and therefore the exosome is termed PM/Scl autoantigen in the autoimmune field. More recently, one of the core components of the exosome was identified as a protein associated with chronic myelogenous leukemia. In this chapter we will describe the identification of the PM/Scl autoantigen from a historical perspective, discuss our current knowledge on the occurrence of autoantibodies to exosome components in autoimmune diseases and end with the data that connect the exosome with cancer.

INTRODUCTION

As described in the other chapters of this book, the exosome is implicated in the processing/maturation and degradation of many different species of RNA. The exosome complex is evolutionary conserved and can therefore be found in virtually all eukaryotic and archaeal forms of life, although the composition of the complex might differ to some extent from one species to another. In eubacteria a similar complex in terms of structure and function can be discerned, which is called the degradosome. The presence of exosome or exosome-like complexes in these different species suggests that it fulfills an essential role in RNA metabolism. In agreement with the assumption that eukaryotic cells can probably not survive without a functional exosome, studies in yeast and humans have shown that an intact exosome core is required for normal cell growth.[1-3]

RNA Exosome, edited by Torben Heick Jensen.

The discovery of the human exosome complex was facilitated by its association with autoimmunity. A response of the immune system to self-components is observed in a plethora of diseases, in particular connective tissue diseases. The autoimmune response can be directed to a variety of biomolecules, including proteins, nucleic acids and lipids. In most cases, it is neither known how immunological tolerance to self-components is broken, nor whether the immune response plays a pathophysiological role or is merely an epiphenomenon. Autoimmunity is generally detected by the appearance of autoantibodies in the serum of patients. Autoantibodies in a distinct group of autoimmune patients target components of the exosome complex and these autoantibodies have been instrumental in the identification of a complex of proteins that was later found to represent the human exosome complex.

IDENTIFICATION OF THE PM/Scl COMPLEX

As already mentioned above, the first indication of the existence of a human exosome complex dates back from 1977 when Wolfe and colleagues showed that autoantibodies in the sera of patients suffering from certain muscle disorders were capable of precipitating an antigen from a calf thymus extract as determined by an immunodiffusion assay.[4] This antigen was initially called "PM-1", as 17 out of the 28 sera (61%) that were scored as 'positive' (in terms of successfully precipitating the antigen) were derived from patients that were diagnosed with polymyositis.

Polymyositis, literally meaning "many muscle inflammations", is a progressive, chronic disorder that belongs to the group of connective tissue diseases (CTD), along with other diseases including dermatomyositis (DM) and systemic sclerosis (SSc, also referred to as scleroderma, Scl). Symptoms of PM include weakening and/or loss of mass of the muscles, which is particularly evident in the legs, shoulders and pelvis of the patient, thereby severely hampering the patient's abilities in everyday's activities such as climbing stairs, standing up or even walking. While some of these PM-symptoms are shared with DM and SSc, the latter two disorders have more prominent visual effects on the skin, such as the appearance of rash, hardening of the skin and disposition of calcium under the skin (Fig. 1). When patients have symptoms of both SSc and PM or DM, the disorder is referred to as polymyositis/scleroderma-overlap syndrome (PM/Scl). A total of 8 patients suffering from this overlap syndrome were also included in Wolfe's study and all but one of these appeared to contain anti-PM-1 antibodies, indicating that these autoantibodies are more common in this group of patients. More importantly however, the disease-specificity of the autoantibodies indicated that they might be exploited as a biomarker for clinical diagnosis, which was not yet available for these patients at that time. In view of the strongest association of the autoantibodies with PM/Scl patients, the antigen was renamed "PM/Scl-antigen" in 1984. In the same period other research groups confirmed the high prevalence of these autoantibodies in patients suffering from the PM/Scl-overlap syndrome.[5,6]

An obvious next step was the identification and characterization of the molecules that make up this particular antigen. By immunoprecipitation experiments it was demonstrated that the PM/Scl-antigen consists of at least 11 polypeptides, with relative molecular masses ranging from 20,000 to 110,000.[7] Subsequently, it took about 5 years before two of the major autoantigenic proteins targeted by the sera of PM/Scl-patients were identified and their cDNAs were cloned.[8-10] The identified proteins were named

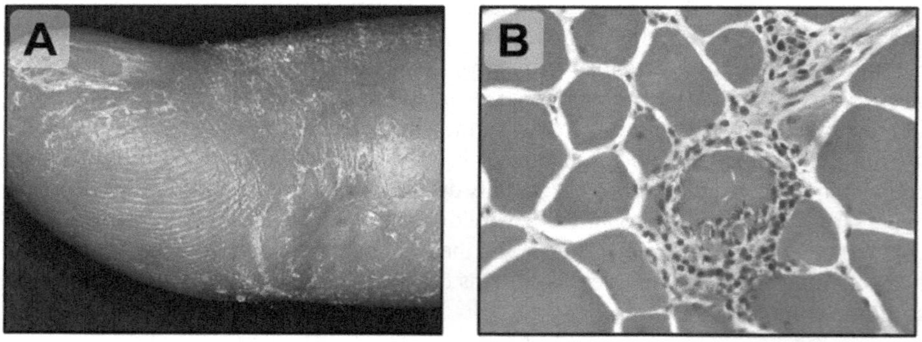

Figure 1. Clinical features of scleroderma and polymyositis. A) Sclerodactyly, a typical symptom often seen in patients suffering from scleroderma, which is characterized by thickening and tightening of the skin. B) Muscle fibers of a polymyositis patient, showing the infiltration of inflammatory cells (a.o. lymphocytes) in the endomysium.

according to their relative molecular mass as determined by SDS-PAGE: a M_r 75,000 protein designated PM/Scl-75 and a M_r 100,000 protein designated PM/Scl-100. It should be noted that the PM/Scl-75 protein migrates aberrantly in SDS-PAGE gels, most likely due to the clustering of charged amino acids in its C-terminal region.[11] In addition, four isoforms of the PM/Scl-75 protein have been described. The original cDNA reported by Alderuccio and coworkers[8] was probably incomplete in the region corresponding to the N-terminus, which was substantiated by the lack of association with the exosome complex in two-hybrid experiments.[12] By screening human EST databases sequences encoding 84 additional amino acids in the N-terminal region were found. Importantly, the polypeptide corresponding to the longer isoform did show two-hybrid interactions with components of the exosome core complex.[11] Another variation in the PM/Scl-75 sequence is resulting from an alternative splicing event, which leads to the incorporation of a 17 amino acids encoding optional exon in the C-terminal region of the protein.

When the yeast exosome complex was identified[3] and Allmang and colleagues found that two of its protein components, Rrp45 and Rrp6, were homologous to the human PM/Scl-75 and PM/Scl-100 proteins, respectively,[2] the suggestion was raised that the PM/Scl complex might in fact represent the human counterpart of the yeast exosome. This was confirmed by the cloning of cDNAs encoding the other components of the human exosome, which was based upon either the homology with their yeast counterparts or the copurification with the human exosome during affinity-purifications.[13,14]

AUTOANTIBODIES TO THE EXOSOME / PM/Scl-ANTIGEN

Originally, the detection of autoantibodies in patient sera was mainly performed by immunodiffusion (or the related technique counterimmunoelectrophoresis) and by indirect immunofluorescence (IF). With the latter technique a typical nucleolar staining pattern was indicative for the presence of anti-PM/Scl autoantibodies and therefore anti-PM/Scl autoantibodies were categorized as anti-nucleolar antibodies (ANoA), together with anti-Th/To (antibodies to protein components of the RNase MRP and RNase P particles)

Table 1. Anti-PM/Scl reactivity in PM, DM, Scl, PM/Scl-overlap and nondifferentiated idiopathic inflammatory myopathy (IIM) patients monitored by ID, IF and/or IP

	IIM		PM		DM		Scl		PM/Scl		
Study	# pos./n	%	# pos./n	%	# pos./n	%	# pos./n	%	# pos./n	%	Method
Wolfe (1977)			9/14	64	1/6	17			7/8	88	ID
Treadwell (1984)			2/22	9			2/32	6	9/77	12	ID
Reichlin (1984)	9/114	8									ID
Reimer (1988)							8/646	1			IF/IP
Reichlin (1988)			8/168	5							ID
Oddis (1992)	5/106	5					6/359	2	10/41	24	ID/IF
Haus-manowa (1997)			0/19	0	0/21	0			19/25	76	ID
O'Hanlon (2006)	65/603	11	13/227	6	19/177	11			32/101	32	ID
Selva-O'Callaghan (2006)	10/88	11	1/27	4	8/59	14					IP
Total	89/911	10	33/477	8	28/263	11	16/1037	2	77/252	31	

and anti-U3 (antibodies to proteins of the U3 snoRNP particle, in particular to fibrillarin). Although this method is not very specific for the detection of anti-PM/Scl autoantibodies, ANoA and ANA (anti-nuclear antibodies) are at least indicative for many CTD, especially since they are usually absent from healthy controls.[15,16] For this reason, this technique was often used as an initial screening method, which was followed by either immunodiffusion (ID), immunoprecipitation (IP), immunoblotting (IB) or enzyme-linked immunosorbent assays (ELISA).

A combination of the data from many studies addressing anti-PM/Scl reactivity in patient sera by either ID, IF and/or IP showed that this reactivity can be found in 31% of PM/Scl patients (Table 1). The frequency of anti-PM/Scl reactivity in patients diagnosed with PM, DM and SSc was 8%, 11% and 2%, respectively.

The cloning of cDNAs encoding individual protein subunits of the exosome and the production of the corresponding recombinant proteins in various expression systems created new possibilities to characterize the anti-PM/Scl autoimmune response and to screen patient sera for the occurrence of anti-PM/Scl autoantibodies. This facilitated the detection of autoantibodies to individual protein components of the exosome, e.g., by ELISA. Since IP assays can be quite laborious, especially when large cohorts of patients have to be analyzed and due to the poor recognition of the main antigens (PM/Scl-75 in particular) by the patient sera on IB, ELISA has become the method of choice to detect

Table 2. Anti-exosome(-associated) protein reactivity in PM, DM, Scl and PM/Scl-overlap patients monitored by ELISA[a]

	Antigen						
Diagnosis	PM/Scl –75	PM/Scl –100	MPP6	C1D	Mtr4	Ski2	hSki8
PM	0–3	2–8	0	5	2	5	0
DM	2–3	2–6	2	0	0	2	2
Scl	10	2–13	0	0	0	0	0
PM/Scl	27–28	23–55	0	23	7	0	3
References	20–22	20–22,24	0	20	0	20	20

[a] Reactivity values are given in percentages.

anti-exosome autoantibodies. During the last decade also a number of exosome-associated proteins, that are not considered to be part of the exosome core complex, but (transiently) associate with a subset of the exosome, have been identified and their cDNAs have been cloned and expressed.[14,17-20] A number of these proteins have been used as well to investigate whether they are also targeted by autoantibodies in (anti-PM/Scl-positive) patient sera (Table 2). A schematic overview of the targeting of exosome core components and of exosome-associated/auxiliary proteins by autoantibodies (as determined by ELISA) in PM/Scl-overlap sera is given in Figure 2.

The majority of anti-PM/Scl-positive patients appeared to have autoantibodies directed against PM/Scl-100 and/or PM/Scl-75 (Table 3). The other exosome core components are less frequently targeted, with the exception of Rrp4, which is recognized by 64% of the anti-PM/Scl-positive patients.[21,22] A combination of ELISA data for PM/Scl-100 and PM/Scl-75 leads to similar sensitivity scores as the conventional ID, IF, IB and IP assays (approximately 31% of PM/Scl patients are positive). After mapping a major autoepitope of PM/Scl-100,[23] a synthetic peptide corresponding to this epitope, designated "PM1-alpha", was produced to set up an ELISA, which allowed the detection of autoantibodies in 55% of the PM/Scl-overlap patients.[24] In addition, the C1D protein, a RNA-binding protein which binds to PM/Scl-100 and participates in exosome-mediated pre-rRNA processing, was found to be a major autoantigen in PM/Scl patients.[19,20]

Table 3. Anti-exosome core protein reactivity in anti-PM/Scl-positive PM, DM, Scl and PM/Scl-overlap patients monitored by ELISA[a]

	Antigen							
Diagnosis	PM/Scl –75	PM/Scl –100	hRrp4	hRrp40	hRrp41	hRrp42	hRrp46	hCsl4
PM	nd	nd	nd	nd	nd	nd	nd	nd
DM	nd	nd	nd	nd	nd	nd	nd	nd
Scl	64	100	64	0	0	36	0	0
PM/Scl	90	100	60	10	0	20	0	0

[a] Reactivity values are given in percentages.

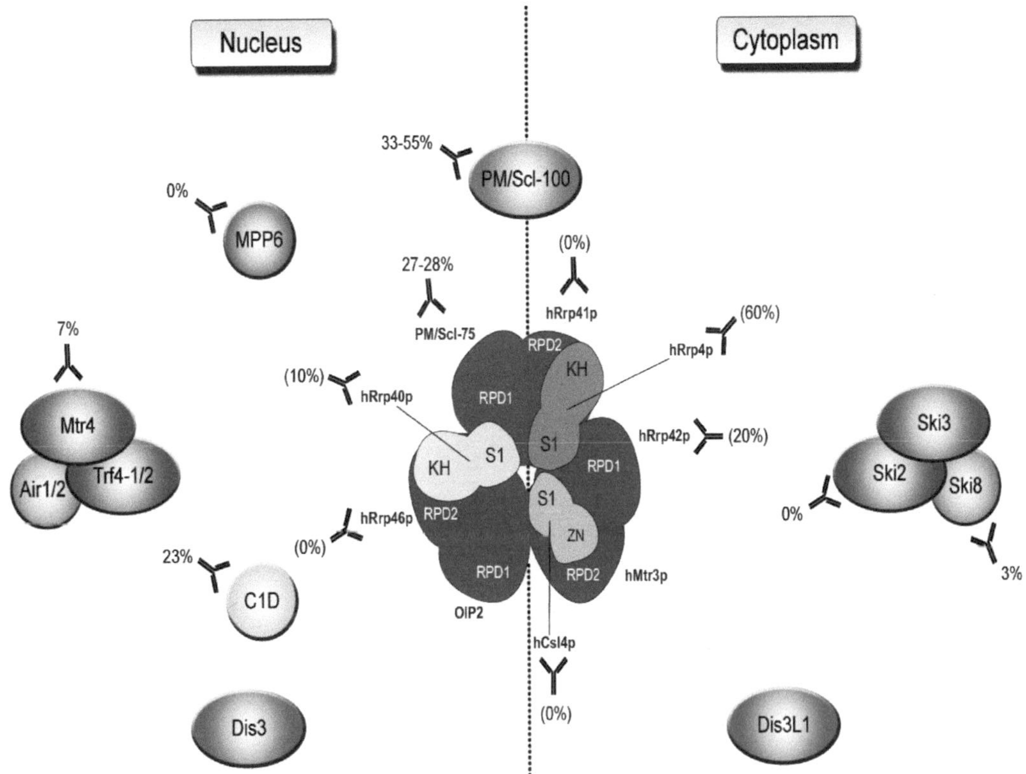

Figure 2. Targeting of exosome and exosome-associated proteins by autoantibodies from PM/Scl-overlap patients. Schematic structure of the human exosome. The nine core subunits assemble into a ring-like structure (central). RPD1 and RPD2 refer to two groups of RNase PH domains; KH, S1 and ZN to RNA-binding domains. Exosome-associated/auxiliary proteins and protein complexes that have been reported to interact with the exosome core are shown at the periphery. Proteins for which the presence of autoantibodies in PM/Scl sera has been investigated are indicated by the schematic antibody structures. Numbers (in percentages) represent the frequency of autoantibodies in patient sera. Numbers between brackcts represent the frequencies for sera that were preselected based upon the reactivity with either PM/Scl-75 or PM/Scl-100.

Because many of the autoantibody studies described above have not yet been replicated with other cohorts of patients, the frequencies by which these antibody specificities occur should be interpreted with care. It is known that ethnic differences and genetic variation may affect the incidence of autoantibody production. The frequency of patients with anti-PM/Scl reactivity, for example, appears to be quite variable, as this reactivity was not found in a large cohort of 275 Japanese patients.[25] This may at least in part be due to the fact that the presence of anti-PM/Scl antibodies seems to be associated with certain MHC alleles, HLA-DRB1*0301, HLA-DQA1*0501 and HLA-DQB1*02,[26,27] whereas HLA-DRB1*15/*16 and HLA-DQA1*0101 might prevent the production of these autoantibodies.[28]

INITIATION OF THE ANTI-EXOSOME / PM/Scl AUTOIMMUNE RESPONSE

It is still an open question what triggers the anti-PM/Scl response in the initial stages of these autoimmune diseases. Several studies have proposed a role for unusual protein modifications, in particular those occurring during apoptosis and necrosis, which might play a role in breaking immunological tolerance to these proteins.[29,30] One of the core subunits of the exosome, PM/Scl-75, which is also one of the most frequently targeted proteins by anti-PM/Scl-positive sera, has been demonstrated to be cleaved by caspases in apoptotic cells.[31] Moreover, PM/Scl-100 appeared to be cleaved by granzyme B in vitro.[32] It is not known yet whether these changes lead to the formation of neo-epitopes. An alternative mechanism might be molecular mimicry, in which structural similarities between epitopes of foreign and self-proteins result in the cross-activation of autoreactive T or B cells by pathogen-derived peptides.

The targeting of multiple exosome subunits and exosome associated proteins is most likely the result of a phenomenon called intermolecular epitope spreading. When an immune response towards a particular protein is elicited, it can extend to another molecule that resides in the same complex. Data supporting this hypothesis regarding the human exosome is still scarce, although two bodies of evidence point in that direction. Firstly, immunization of rabbits with a synthetic peptide corresponding to a major epitope of PM/Scl-100 led to the generation of antibodies that targeted other components of the exosome complex as well.[33] Secondly, the serum of a patient with high anti-PM/Scl reactivity was shown to stain a single, 100 kDa polypeptide in IB. Three months later, the serum was found to stain an additional, 29 kDa polypeptide in IB, which was suggested to correspond to an aberrant form of PM/Scl-75, but might be one of the other core components as well.[34]

THE HUMAN EXOSOME AND CANCER

As already mentioned in the introduction, the viability of cells is severely impaired when the integrity of the exosome complex is disturbed. This strongly suggests that at least one of the functions of the exosome is essential to keep the cell in a proliferating state. On the other hand, when the activity of the exosome would be rate limiting in the maturation or turnover of RNAs that are crucial for cell proliferation, an overactive exosome may lead to a higher rate of proliferation. Although rather speculative, this is supported by observations of Yang and coworkers who examined patients who received a donor lymphocyte infusion (DLI) as a treatment for chronic myelogenous leukemia (CML).[35] CML is characterized by the uncontrolled production of myeloid cells, resulting in a strong accumulation of these cells in the blood and is associated with a typical chromosomal translocation resulting in the so-called Philadelphia chromosome. This translocation leads to the production of an aberrant, overactive tyrosine kinase fusion protein, Bcr-Abl, which is thought to affect several substrate proteins involved in cell division, DNA repair and genomic instability. Of all the known cases of adult leukemia in the Western countries, about 15 to 20% have been classified as CML.[36] Before the introduction of Imatinib, a drug that specifically counteracts the activity of the Bcr-Abl fusion protein, patients suffering from CML often received a bone marrow transplantation (BMT). The patients who relapsed after this transplantation were further treated with DLI, resulting in a durable remission of the disease in 70 to 80% of the cases.[37] The success of this treatment is a result of a phenomenon called the graft-versus-tumor (GVT) effect, in which the infused lymphocytes attack any remaining cancerous cells in the bone marrow.

When Yang and colleagues used the serum of DLI-responding patients for an antibody-based screening of a CML cDNA expression library, new putative tumor-related antigens were identified.[38] Among these was a M_r 28,000 protein, which was designated CML28. Sequence analysis revealed that this protein is identical to hRrp46, a constituent of the exosome core complex (Fig. 2), although the 5' coding region of the CML28 cDNA is 33 amino acids longer than that of hRrp46. To confirm that hRrp46 is immunogenic in these patients, serum samples derived from a CML patient before and after BMT and DLI were analyzed with IB and ELISA using recombinant hRrp46. While anti-hRrp46 reactivity couldn't be detected in healthy controls and in the CML patient prior to DLI, this reactivity significantly increased 2 to 6 months post-DLI, after which the reactivity gradually declined to undetectable levels 2 years post-DLI. Moreover, the temporal pattern of anti-hRrp46 reactivity correlated well with the onset of cytogenetic remission, which is indicated by the disappearance of the Philadelphia chromosome. Since the patient didn't appear to have any symptoms indicating the development of an autoimmune disease, the anti-hRrp46 reactivity seemed to be associated with tumor rejection instead. The immune response towards hRrp46 was shown to be a common feature of other types of cancer, such as lung and prostate cancer and melanoma, in which the antibodies could be detected in 10% to 33% of the cases. The anti-hRrp46 response might at least in part be due to the overexpression of hRrp46 in these different cancers, as it was demonstrated by northern blot hybridization and IB that highly proliferating cell lines express high levels of hRrp46 when compared to normal tissues or even stable-phase CML. These findings might also prove useful for the development of an antigen-specific immunotherapy and progress in the development of such strategies has already been made.[39-41] The anti-hRrp46 immune response and elevated hRrp46 expression levels in cancer raise the question whether these phenomena are specific for hRrp46 or also occur for other exosome components. Further experiments will be required to clarify this issue.

Interestingly, the antimetabolite 5-fluorouracil, which is frequently used to treat solid tumors in a variety of cancers, was shown to inhibit the function of the exosome. Originally this drug was selected for its hypothetical ability to inhibit cell proliferation, presumably by causing thymidine starvation and thus negatively affecting DNA synthesis. However, Fang and coworkers reported that also rRNA processing was impaired by this drug.[42] The accumulated pre-rRNA precursors were similar to the precursors accumulating in exosome mutant yeast strains. Possibly related to these observations, it was demonstrated that the absence of Rrp6 (the yeast counterpart of PM/Scl-100) enhanced the 5-fluorouracil-induced defects.[43] Taken together, these results suggest that 5-fluorouracil exerts its effect on rRNA processing at least in part by inhibiting the exosome.

CONCLUSION

In view of the central role of the human exosome in the processing and degradation of many RNAs, it is intriguing to investigate to what extent a disturbance of its function can interfere with normal cellular physiology and is associated with diseases. The two best documented examples, autoimmunity and cancer, have been discussed in this chapter, although there are still many questions that need to be addressed. It is, for example, still unclear what causes the targeting of PM/Scl-75 and PM/Scl-100 by the immune system in patients suffering from the PM/Scl-overlap syndrome and why this response is particularly associated with this autoimmune disease. In view of the physical barrier of the plasma membrane, autoantibodies are not likely to interfere with the functions of the exosome in RNA metabolism, but once exosome components are released from (dying) cells the resulting immune complexes may contribute to the progression of autoimmunity and tissue damage. In case of cancer dysfunctioning of the exosome may play a direct role, but it is clear that more work has to be done to investigate the effects of increased or decreased exosome activity on cell proliferation. It is currently not known whether exosome core subunits other than hRrp46 are overexpressed in solid tumors as well. Alternatively, a pool of non-exosome associated hRrp46 may exist and might have a completely other role independent of the exosome. In this respect, also the importance of the N-terminal extension, which may be specific for a particular isoform of hRrp46, needs to be studied.

The exosome is known to play an important role in the degradation of a special class of mRNAs, containing cis-acting adenylate-uridylate-rich sequence elements (AREs) in their 3′UTR. Transcripts containing such elements are often involved in important biological processes and their levels are therefore often tightly regulated. In theory, a perturbation of exosome function is likely to affect these levels and, as a consequence, might cause a wide variety of pathological conditions.[44]

In this chapter we focused on human diseases, but one should realize that the exosome is expressed in many other species and thus might also here be associated with disease. In plants, for instance, it was found that the effects of deletion of a particular exosome component mimicked a disorder that induced cell death in the tip of the leafs when plants are inoculated with *Blumeria graminis*, a fungus that causes mildew on grasses and that is frequently used to infect plants in a laboratory setting. Intriguingly, the deleted exosome component was Rrp46.[45]

We conclude that although targeting of exosome components in autoimmunity is known for many years, our knowledge on the involvement of the exosome, of its individual components or of its auxiliary factors in other diseases is still in its infancy and much work needs to be done to obtain more insight in this topic.

ACKNOWLEDGEMENTS

We thank Reinout Raijmakers as well as present and former members of our laboratory for stimulating discussions and contribution of ideas and information to this chapter.

REFERENCES

1. van Dijk EL, Schilders G, Pruijn GJ. Human cell growth requires a functional cytoplasmic exosome, which is involved in various mRNA decay pathways. RNA 2007, 13:1027-1035.
2. Allmang C, Petfalski E, Podtelejnikov A et al. The yeast exosome and human PM-Scl are related complexes of 3′ —> 5′ exonucleases. Genes Dev 1999; 13:2148-2158.
3. Mitchell P, Petfalski E, Shevchenko A et al. The exosome: a conserved eukaryotic RNA processing complex containing multiple 3′—>5′ exoribonucleases. Cell 1997; 91:457-466.
4. Wolfe JF, Adelstein E, Sharp GC. Antinuclear antibody with distinct specificity for polymyositis. J Clin Invest 1977; 59:176-178.
5. Treadwell EL, Alspaugh MA, Wolfe JF et al. Clinical relevance of PM-1 antibody and physiochemical characterization of PM-1 antigen. J Rheumatol 1984; 11:658-662.
6. Reichlin M, Maddison PJ, Targoff I et al. Antibodies to a nuclear/nucleolar antigen in patients with polymyositis overlap syndromes. J Clin Immunol 1984; 4:40-44.
7. Reimer G, Scheer U, Peters JM et al. Immunolocalization and partial characterization of a nucleolar autoantigen (PM-Scl) associated with polymyositis/scleroderma overlap syndromes. J Immunol 1986; 137:3802-3808.
8. Alderuccio F, Chan EK, Tan EM. Molecular characterization of an autoantigen of PM-Scl in the polymyositis/scleroderma overlap syndrome: a unique and complete human cDNA encoding an apparent 75-kD acidic protein of the nucleolar complex. J Exp Med 1991; 173:941-952.
9. Bluthner M, Bautz FA. Cloning and characterization of the cDNA coding for a polymyositis-scleroderma overlap syndrome-related nucleolar 100-kD protein. J Exp Med 1992; 176:973-980.
10. Ge Q, Frank MB, O'Brien C et al. Cloning of a complementary DNA coding for the 100-kD antigenic protein of the PM-Scl autoantigen. J Clin Invest 1992; 90:559-570.
11. Raijmakers R, Vree Egberts WT, van Venrooij WJ et al. The association of the human PM/Scl-75 autoantigen with the exosome is dependent on a newly identified N terminus. J Biol Chem 2003; 278:30698-30704.
12. Raijmakers R, Vree Egberts WT, van Venrooij WJ et al. Protein-protein interactions between human exosome components support the assembly of RNase PH-type subunits into a six-membered PNPase-like ring. J Mol Biol 2002; 323:653-663.
13. Brouwer R, Allmang C, Raijmakers R et al. Three novel components of the human exosome. J Biol Chem 2001; 276:6177-6184.
14. Chen CY, Gherzi R, Ong SE et al. AU binding proteins recruit the exosome to degrade ARE-containing mRNAs. Cell 2001; 107:451-464.
15. von Muhlen CA, Tan EM. Autoantibodies in the diagnosis of systemic rheumatic diseases. Semin Arthritis Rheum 1995; 24:323-358.
16. Reimer G, Steen VD, Penning CA et al. Correlates between autoantibodies to nucleolar antigens and clinical features in patients with systemic sclerosis (scleroderma). Arthritis Rheum 1988; 31:525-532.
17. Lehner B, Sanderson CM. A protein interaction framework for human mRNA degradation. Genome Res 2004; 14:1315-1323.
18. Schilders G, Raijmakers R, Raats JM et al. MPP6 is an exosome-associated RNA-binding protein involved in 5.8S rRNA maturation. Nucleic Acids Res 2005; 33:6795-6804.
19. Schilders G, van Dijk EL, Pruijn GJ. C1D and hMtr4p associate with the human exosome subunit PM/Scl-100 and are involved in prerRNA processing. Nucleic Acids Res 2007; 35:2564-2572.
20. Schilders G, Vree Egberts WT, Raijmakers R et al. C1D is a major autoantibody target in patients with the polymyositis-scleroderma overlap syndrome. Arthritis Rheum 2007; 56:2449-2454.
21. Brouwer R, Vree Egberts WT, Hengstman GJ et al. Autoantibodies directed to novel components of the PM/Scl complex, the human exosome. Arthritis Res 2002; 4:134-138.
22. Raijmakers R, Renz M, Wiemann C et al. PM-Scl-75 is the main autoantigen in patients with the polymyositis/scleroderma overlap syndrome. Arthritis Rheum 2004; 50:565-569.
23. Bluthner M, Mahler M, Muller DB et al. Identification of an alpha-helical epitope region on the PM/Scl-100 autoantigen with structural homology to a region on the heterochromatin p25beta autoantigen using immobilized overlapping synthetic peptides. J Mol Med 2000; 78:47-54.
24. Mahler M, Raijmakers R, Dahnrich C et al. Clinical evaluation of autoantibodies to a novel PM/Scl peptide antigen. Arthritis Res Ther 2005; 7:R704-R713.
25. Kuwana M, Okano Y, Kaburaki J et al. Racial differences in the distribution of systemic sclerosis-related serum antinuclear antibodies. Arthritis Rheum 1994; 37:902-906.
26. Hausmanowa-Petrusewicz I, Kowalska-Oledzka E, Miller FW et al. Clinical, serologic and immunogenetic features in Polish patients with idiopathic inflammatory myopathies. Arthritis Rheum 1997; 40:1257-1266.

27. Chinoy H, Salway F, Fertig N et al. In adult onset myositis, the presence of interstitial lung disease and myositis specific/associated antibodies are governed by HLA class II haplotype, rather than by myositis subtype. Arthritis Res Ther 2006; 8:R13.
28. O'Hanlon TP, Carrick DM, Targoff IN et al. Immunogenetic risk and protective factors for the idiopathic inflammatory myopathies: distinct HLA-A, -B, -Cw, -DRB1 and -DQA1 allelic profiles distinguish European American patients with different myositis autoantibodies. Medicine (Baltimore) 2006; 85:111-127.
29. Utz PJ, Gensler TJ, Anderson P. Death, autoantigen modifications and tolerance. Arthritis Res 2000; 2:101-114.
30. Hof D, Pruijn GJM, van Venrooij WJ et al. The role of cell death-specific modifications in breaking tolerance to self-antigens. In: Kettleworth CR, ed. Cell Apoptosis Research Advances. Hauppauge: Nova Science Publishers, 2007:179-202.
31. Schilders G, Raijmakers R, Malmegrim KC et al. Caspase-mediated cleavage of the exosome subunit PM/Scl-75 during apoptosis. Arthritis Res Ther 2007; 9:R12.
32. Casciola-Rosen L, Andrade F, Ulanet D et al. Cleavage by granzyme B is strongly predictive of autoantigen status: implications for initiation of autoimmunity. J Exp Med 1999; 190:815-826.
33. Mahler M, Raijmakers R. Novel aspects of autoantibodies to the PM/Scl complex: clinical, genetic and diagnostic insights. Autoimmun Rev 2007; 6:432-437.
34. Gutierrez-Ramos R, Gonz Lez-Diaz V, PachecoTovar MG et al. A dermatomyositis and scleroderma overlap syndrome with a remarkable high titer of anti-exosome antibodies. Reumatismo 2008; 60:296-300.
35. Yang XF, Wu CJ, Chen L et al. CML28 is a broadly immunogenic antigen, which is overexpressed in tumor cells. Cancer Res 2002; 62:5517-5522.
36. Faderl S, Talpaz M, Estrov Z et al. Chronic myelogenous leukemia: biology and therapy. Ann Intern Med 1999; 131:207-219.
37. Collins RH, Jr., Shpilberg O, Drobyski WR et al. Donor leukocyte infusions in 140 patients with relapsed malignancy after allogeneic bone marrow transplantation. J Clin Oncol 1997; 15:433-444.
38. Wu CJ, Yang XF, McLaughlin S et al. Detection of a potent humoral response associated with immune-induced remission of chronic myelogenous leukemia. J Clin Invest 2000; 106:705-714.
39. Zhou H, Zhang D, Wang Y et al. Induction of CML28-specific cytotoxic T-cell responses using cotransfected dendritic cells with CML28 DNA vaccine and SOCS1 small interfering RNA expression vector. Biochem Biophys Res Commun 2006; 347:200-207.
40. Xie LH, Sin FW, Cheng SC et al. Activation of cytotoxic T lymphocytes against CML28-bearing tumors by dendritic cells transduced with a recombinant adeno-associated virus encoding the CML28 gene. Cancer Immunol Immunother 2008; 57:1029-1038.
41. Mao H, Geng Z, Liu W et al. Selection of HLA-A2 restricted CML28 peptide by artificial antigen-presenting cells. J Immunother 2008; 31:487-490.
42. Fang F, Hoskins J, Butler JS. 5-fluorouracil enhances exosome-dependent accumulation of polyadenylated rRNAs. Mol Cell Biol 2004; 24:10766-10776.
43. Lum PY, Armour CD, Stepaniants SB et al. Discovering modes of action for therapeutic compounds using a genome-wide screen of yeast heterozygotes. Cell 2004; 116:121-137.
44. Khabar KS. The AU-rich transcriptome: more than interferons and cytokines and its role in disease. J Interferon Cytokine Res 2005; 25:1-10.
45. Xi L, Moscou MJ, Meng Y et al. Transcript-based cloning of RRP46, a regulator of rRNA processing and R gene-independent cell death in barley-powdery mildew interactions. Plant Cell 2009; 21:3280-3295.

INDEX

R

S

T

V

X

Y